LIFE SCIENCE

Answer Key & Parent Companion

Science
Shepherd

SCIENCE SHEPHERD
LIFE SCIENCE ANSWER KEY & PARENT COMPANION

Published by:
Ohana Life Press, LLC
1405 Capitol Dr.
Suite C-202
Pewaukee, WI 53072
www.ScienceShepherd.com

Written by Scott Hardin, MD

ACKNOWLEDGEMENTS:

Cover Design
Alex Hardin

Photo Credits
iStock.com

Copy Editing
Stacie O'Brien

Graphic Design
Jason Brown

ISBN: 978-0-9814587-6-2

Table of Contents

Introduction

The purpose of this booklet is two-fold; first, it contains the answers to all of the test and study questions that you (or your student) can use to grade tests and check progress on each chapter. Second, the Parent Companion section provides brief summaries of every chapter's sub-sections in the textbook. The summarized information is designed to give you rapid access to the main concepts your student is studying. In addition, you will also find one or two questions with answers that you can ask your student, if you are so inclined. Even if you know nothing about Life Science, since the answers are also provided with the questions, you can use this section to ask specific questions of your student to see how they are progressing and grasping new concepts. Many parents use the summarized material and questions in the Companion to help them stay connected to what their child is learning by asking them one or more questions every now and then. However, it is totally optional and other parents do not use this section at all. Please take a little time and familiarize yourself with the Companion to see if it will fit into your homeschool!

Schedule

WEEK 1	DAY ONE	DAY TWO	DAY THREE	DAY FOUR	DAY FIVE
TEXTBOOK	Sections 1.0-1.2	Sections 1.3-1.4	Sections 1.5-1.6	Chapter 1 Study Questions	Sections 2.0-2.3
TEST BOOKLET					

WEEK 2	DAY ONE	DAY TWO	DAY THREE	DAY FOUR	DAY FIVE
TEXTBOOK	Sections 2.4-2.6	Section 2.7	Sections 2.8-2.10	Sections 2.11-2.13	Chapter 2 Study Questions
TEST BOOKLET					

WEEK 3	DAY ONE	DAY TWO	DAY THREE	DAY FOUR	DAY FIVE
TEXTBOOK	Review Chapters 1 & 2	Sections 3.0-3.3	Sections 3.4-3.6	Sections 3.7-3.8	Section 3.9
TEST BOOKLET					

WEEK 4	DAY ONE	DAY TWO	DAY THREE	DAY FOUR	DAY FIVE
TEXTBOOK	Sections 3.10-3.11	Chapter 3 Study Questions	Review Chapters 1-3		Sections 4.0-4.2
TEST BOOKLET				Test #1	

WEEK 5	DAY ONE	DAY TWO	DAY THREE	DAY FOUR	DAY FIVE
TEXTBOOK	Sections 4.3-4.4	Section 4.5	Section 4.6	Sections 4.7-4.8	Sections 4.9-4.10
TEST BOOKLET					

WEEK 6	DAY ONE	DAY TWO	DAY THREE	DAY FOUR	DAY FIVE
TEXTBOOK	Section 4.11	Review Chapter 4	Chapter 4 Study Questions	Sections 5.0-5.3	Sections 5.4-5.6
TEST BOOKLET					

WEEK 7	DAY ONE	DAY TWO	DAY THREE	DAY FOUR	DAY FIVE
TEXTBOOK	Sections 5.7-5.9	Sections 5.10-5.13	Section 5.14	Section 5.15	Sections 5.16-5.17
TEST BOOKLET					

WEEK 8	DAY ONE	DAY TWO	DAY THREE	DAY FOUR	DAY FIVE
TEXTBOOK	Section 5.18	Review Chapter 5	Chapter 5 Study Questions	Review Chapters 4 & 5	
TEST BOOKLET					Test #2

WEEK 9	DAY ONE	DAY TWO	DAY THREE	DAY FOUR	DAY FIVE
TEXTBOOK	Sections 6.0-6.2	Section 6.3	Section 6.4	Sections 6.5-6.6	Sections 6.7-6.8
TEST BOOKLET					

WEEK 10	DAY ONE	DAY TWO	DAY THREE	DAY FOUR	DAY FIVE
TEXTBOOK	Sections 6.9-6.10	Sections 6.11-6.12	Sections 6.13-6.14	Review Chapter 6	Chapter 6 Study Questions
TEST BOOKLET					

WEEK 11	DAY ONE	DAY TWO	DAY THREE	DAY FOUR	DAY FIVE
TEXTBOOK	Sections 7.0-7.2	Sections 7.3-7.4	Section 7.5	Section 7.6	Section 7.7
TEST BOOKLET					

WEEK 12	DAY ONE	DAY TWO	DAY THREE	DAY FOUR	DAY FIVE
TEXTBOOK	Sections 7.8-7.9	Review Chapter 7	Chapter 7 Study Questions	Review Chapters 6 & 7	
TEST BOOKLET					Test #3

WEEK 13	DAY ONE	DAY TWO	DAY THREE	DAY FOUR	DAY FIVE
TEXTBOOK	Sections 8.0-8.2	Section 8.3	Sections 8.4-8.5	Sections 8.6-8.7	Sections 8.8-8.9
TEST BOOKLET					

WEEK 14	DAY ONE	DAY TWO	DAY THREE	DAY FOUR	DAY FIVE
TEXTBOOK	Sections 8.10-8.12	Sections 8.13-8.16	Section 8.17	Sections 8.18-8.19	Review Chapter 8
TEST BOOKLET					

WEEK 15	DAY ONE	DAY TWO	DAY THREE	DAY FOUR	DAY FIVE
TEXTBOOK	Chapter 8 Study Questions	Sections 9.0-9.2	Sections 9.3-9.4	Sections 9.5-9.6	Section 9.7
TEST BOOKLET					

WEEK 16	DAY ONE	DAY TWO	DAY THREE	DAY FOUR	DAY FIVE
TEXTBOOK	Section 9.8	Sections 9.9-9.10	Section 9.11	Section 9.12	Section 9.13
TEST BOOKLET					

WEEK 17	DAY ONE	DAY TWO	DAY THREE	DAY FOUR	DAY FIVE
TEXTBOOK	Sections 9.14-9.16	Review Chapter 9	Chapter 9 Study Questions	Review Chapters 8 & 9	
TEST BOOKLET					Test #4

WEEK 18	DAY ONE	DAY TWO	DAY THREE	DAY FOUR	DAY FIVE
TEXTBOOK	Sections 10.0-10.1	Section 10.2	Sections 10.3	Section 10.4	Section 10.5
TEST BOOKLET					

WEEK 19	DAY ONE	DAY TWO	DAY THREE	DAY FOUR	DAY FIVE
TEXTBOOK	Sections 10.6-10.7	Review Chapter 10	Chapter 10 Study Questions	Sections 11.0-11.1	Sections 11.2-11.3
TEST BOOKLET					

WEEK 20	DAY ONE	DAY TWO	DAY THREE	DAY FOUR	DAY FIVE
TEXTBOOK	Sections 11.4-11.6	Sections 11.7-11.8	Sections 11.9-11.10	Sections 11.11-11.12	Sections 11.13-11.14
TEST BOOKLET					

WEEK 21	DAY ONE	DAY TWO	DAY THREE	DAY FOUR	DAY FIVE
TEXTBOOK	Review Chapter 11	Chapter 11 Study Questions	Review Chapters 10 & 11		Sections 12.0-12.2
TEST BOOKLET				Test #5	

WEEK 22	DAY ONE	DAY TWO	DAY THREE	DAY FOUR	DAY FIVE
TEXTBOOK	Sections 12.3-12.5	Sections 12.6-12.7	Sections 12.8-12.9	Sections 12.10-12.12	Sections 12.13-12.15
TEST BOOKLET					

WEEK 23	DAY ONE	DAY TWO	DAY THREE	DAY FOUR	DAY FIVE
TEXTBOOK	Sections 12.16-12.18	Review Chapter 12	Chapter 12 Study Questions	Sections 13.0-13.3	Sections 13.4-13.5
TEST BOOKLET					

WEEK 24	DAY ONE	DAY TWO	DAY THREE	DAY FOUR	DAY FIVE
TEXTBOOK	Sections 13.6-13.7	Sections 13.8-13.11	Sections 13.12-13.16	Sections 13.17-13.18	Review Chapter 13
TEST BOOKLET					

WEEK 25	DAY ONE	DAY TWO	DAY THREE	DAY FOUR	DAY FIVE
TEXTBOOK	Chapter 13 Study Questions	Review Chapters 12 & 13		Sections 14.0-14.2	Sections 14.3-14.4
TEST BOOKLET			Test #6		

WEEK 26	DAY ONE	DAY TWO	DAY THREE	DAY FOUR	DAY FIVE
TEXTBOOK	Sections 14.5-14.7	Sections 14.8-14.9	Sections 14.10-14.11	Sections 14.12-14.14	Review Chapter 14
TEST BOOKLET					

WEEK 27	DAY ONE	DAY TWO	DAY THREE	DAY FOUR	DAY FIVE
TEXTBOOK	Chapter 14 Study Questions	Sections 15.0-15.1	Sections 15.2-15.3	Section 15.4	Section 15.5
TEST BOOKLET					

WEEK 28	DAY ONE	DAY TWO	DAY THREE	DAY FOUR	DAY FIVE
TEXTBOOK	Section 15.6	Section 15.7	Sections 15.8-15.9	Review Chapter 15	Chapter 15 Study Questions
TEST BOOKLET					

WEEK 29	DAY ONE	DAY TWO	DAY THREE	DAY FOUR	DAY FIVE
TEXTBOOK	Sections 16.0-16.1	Section 16.2	Section 16.3	Section 16.4	Sections 16.5-16.6
TEST BOOKLET					

WEEK 30	DAY ONE	DAY TWO	DAY THREE	DAY FOUR	DAY FIVE
TEXTBOOK	Review Chapter 16	Chapter 16 Study Questions	Review Chapters 15 & 16		Sections 17.0-17.1
TEST BOOKLET				Test #7	

WEEK 31	DAY ONE	DAY TWO	DAY THREE	DAY FOUR	DAY FIVE
TEXTBOOK	Sections 17.2-17.4	Section 17.5	Section 17.6	Sections 17.7-17.9	Sections 17.10-17.12
TEST BOOKLET					

WEEK 32	DAY ONE	DAY TWO	DAY THREE	DAY FOUR	DAY FIVE
TEXTBOOK	Sections 17.13-17.14	Sections 17.15-17.17	Section 17.18	Section 17.19	Sections 17.20-17.21
TEST BOOKLET					

WEEK 33	DAY ONE	DAY TWO	DAY THREE	DAY FOUR	DAY FIVE
TEXTBOOK	Review Chapter 17	Chapter 17 Study Questions	Sections 18.0-18.2	Section 18.3	Section 18.4
TEST BOOKLET					

WEEK 34	DAY ONE	DAY TWO	DAY THREE	DAY FOUR	DAY FIVE
TEXTBOOK	Sections 18.5-18.7	Sections 18.8-18.9	Section 18.10	Sections 18.11-18.12	Sections 18.13-18.16
TEST BOOKLET					

WEEK 35	DAY ONE	DAY TWO	DAY THREE	DAY FOUR	DAY FIVE
TEXTBOOK	Review Chapter 18	Chapter 18 Study Questions	Review Chapters 17 & 18		Sections 19.0-19.4
TEST BOOKLET				Test #8	

WEEK 36	DAY ONE	DAY TWO	DAY THREE	DAY FOUR	DAY FIVE
TEXTBOOK	Sections 19.5-19.7	Sections 19.8-19.12	Review Chapter 19	Chapter 19 Study Questions	
TEST BOOKLET					Test #9

Study Question Answer Key

CHAPTER 1

1. Life science is the study of living organisms. Biology is another word for life science. People who study life science are called biologists.

2. The scientific method is the systematic process by which scientists ask and then answer questions.

   The scientific method begins with the scientist identifying the question which they are interested in answering. Data and observations are gathered in an attempt to better understand the question being asked.

   Once data and observations have been made in the initial phase, a hypothesis statement is made. The hypothesis statement attempts to explain the data gathered up to that point and answer a specific question. It also attempts to predict future data regarding the question.

   Following the hypothesis statement the scientist then devises experiments in order to test whether or not the hypothesis statement holds up to scientific testing. If the scientist finds that the hypothesis is supported by experiments, then further experimentation is performed.

   If the scientist finds that the data does not support the hypothesis statement, then the hypothesis statement is either reformulated or is thrown out entirely and a new one is made.

   Hypothesis statements, which withstand rigorous experimental testing over a number of years, become theories. Theories which withstand further testing, become laws.

3. The data collection phase is called observation and experimentation.

4. A controlled experiment is an experiment designed to test the effects of changing one variable or factor upon two groups which are being studied.

5. A theory is a hypothesis statement that has held up repeatedly to scientific studies.

6. SI units stands for, 'the international system of measurments'. They are helpful for scientific studies because they represent universally accepted standards of measurement that all scientists understand.

7. Five senses

   Instruments such as thermometers, speedometers, tape measures, CT scanners, MRI, microscopes, etc.

   Computers

CHAPTER 2

1. All living things: Are composed of one or more cells. Contain DNA. Reproduce either sexually, asexually, or both. Are both complex and highly organized. Have a way to sense and respond to changes in their environment. Have a way to exract energy from their surroundings. Have a way to maintain homeostasis.

2. A unicellular organism is composed of one cell. A multicellular organism is composed of two or more cells

3. DNA contains the information that every cell and every organism needs in order to function, live, and reproduce properly.

4. There are many levels of complexity and organization displayed by living organisms. Organisms are structurally and biochemically complex and organized, controlled by their highly organized and complex DNA. The interactions of organisms with one another and their environment also displays this property.

5. A receptor is a structure that an organism has in order to sense changes in its environment.

6. Photosynthesis is the process by which plants convert the sun's energy into food that they can use. Photosythesis is important for a number of reasons: It is the main way that the oxygen in our air is replenished. It is the starting point for all energy on earth.

7. An herbivore's food source is plants. A carnivore's food source is other animals or other consumers.

8. Cellular respiration is the process by which cells convert the food that they eat or make, into usable energy for the cell or the organism.

9. The maintenance of a stable environment inside of an organism or cell is called homeostasis.

10. Taxonomy is the scientific process for classifying all species or all organisms into groups.

11. All organisms require: food, a habitat, water
    Almost all organisms require: food, a habitat, water, oxygen

12. False.

CHAPTER 3

1. Anything that has mass and takes up space is called matter.
2. False.
3. An atom is composed of: Neutrons Protons Electrons
4. Protons are found in the nucleus. Electrons are found outside of the nucleus.
5. An element is a molecule composed of only one type of atom.
6. An organic molecule is a molecule that contains carbon, and is made by living organisms. The four main classes of organic molecules are: Carbohydrates, Lipids, Proteins, Nucleic acids
7. A simple carbohydrate is composed of 1, 2, or 3 individual carbohydrate molecules linked together. A complex carbohydrate is composed of 4 or more carbohydrate molecules linked together.
8. True.
9. The majority of the structure of all living organisms is composed of lipids and proteins.
10. False. Meat is a good source of protein, potatoes are not.
11. Sunlight is the ultimate source of energy.
12. The EMS is made up of gamma rays, x rays, ultraviolet rays, visible light, infrared light, microwaves and radio waves.
13. The light is broken into the visible spectrum.
14. False.
15. False.

CHAPTER 4

1. A basic cell structure is a cell membrane enclosing cytoplasm and organelles.

2. Cell theory states that the cell is the individual functional unit of life.

3. All cells contain DNA. All cells use energy. All cells reproduce. All cells grow. All cells maintain homeostasis. All cells respond to their environments. All cells are complex and organized.

4. An organ is a group of individual tissues, which have or share a common function.

5. The outer boundary of a cell is the cell membrane. The outer boundary of a cell that has a cell wall, is the cell wall.

6. Eukaryotic organelles are membrane bound, organelles are not. The DNA of eukaryotic cells is in the nucleus, DNA of prokaryotic cells is free in the cytoplasm, in an area called the nucleoid. All prokaryotic organisms are single-celled. Eukaryotic organisms can be single-cell or multi-cell.

7. Peptidoglycan. Cellulose. Chitin.

8. The lipid bilayer is the name of the double layer of lipids that makes up the cell membrane. The individual molecules that form the lipid bilayer are called phospholipid molecules

9. Osmosis.

   Diffusion.

   Passive transport.

10. Active transport.

    Endocytosis.

    Exocytosis.

CHAPTER 5

1. Both prokaryotic and eukaryotic cells contain the ribosome.

2. None.

3. Endoplasmic reticulum.

4. Ribosomes make proteins.

5. The nucleolus manufactures components or pieces of ribosomes.

6. The cell wall is found outside of the cell membrane and is an added layer of protection and support. Cell walls are found in plants, fungi, algae and bacteria. Animal cells do NOT have cell walls.

7. False.

8. Aerobic cellular respiration occurs inside the mitochondria. During this process, glucose is metabolized (broken down) and the released energy is used to make ATP.

9. Photosynthesis occurs in the chloroplast. During this process, energy from sunlight is captured and used to fuel the reactions to make glucose.

10. False.

11. True.

12. True.

13. Anaerobic cellular respiration is able to supply ATP more quickly than aerobic cellular respiration can.

CHAPTER 6

1. Adenine, thymine, guanine and cytosine.

2. The basic structure of a DNA nucleotide is deoxyribose. With a phosphate group attached to one end, and a nitrogen base attached to the other.

3. False.

4. DNA is cut into smaller units called chromosomes, and each chromosome is organized into even smaller units called genes.

5. A gene is a segment of DNA that codes for the production of one protein. A codon is a three nucleotide segment of a gene that codes for the insertion of one amino acid into a protein.

6. Transcription is the process of making a molecule of mRNA from a DNA template.

7. Translation is the process of a ribosome making or synthesizing a protein molecule by reading the codons of an mRNA molecule.

8. The ribosome knows that the tRNA molecule is carrying the proper amino acid if the three nucleotide sequence on the tRNA molecule, is complementary to the codon of the mRNA.

9. DNA replication is necessary because of the process of cellular division.

10. A mutation is the change in the normal sequence of DNA.

11. True.

12. False.

CHAPTER 7

1. Binary fission is the asexual cellular reproduction process that is performed by prokaryotic cells.

2. Asexual reproduction results in the formation of diploid daughter cells from a diploid parent cell.

3. By definition, a diploid cell contains two pairs of every chromosome.

4. By definition, a haploid cell contains one pair of every chromosome.

5. You would say that the organism was budding.

6. A zygote is the new cell, or the new organism, that is as a result of sexual reproduction.

7. Gametes can't be formed through mitosis, because the zygote would contain too many chromosomes.

8. During meiosis, there are two events of cell division.

9. False.

CHAPTER 8

1. Heredity is the passage of traits (or genes, alleles, or characteristics) from generation to generation.

2. Genetics is the study of heredity.

3. Pollen from one flower is transferred to another flower.

4. Each one of the F1 offspring plants inherited one dominant allele from the YY parent plant. Therefore, each F1 plant contained one dominant allele for pea color. The presence of the dominant allele suppressed the trait of the recessive allele and so all peas had the trait of the dominant allele, or were yellow in color.

5. The F1 organisms were all hybrids, or contained one dominant and one recessive allele. The recessive trait (green pea color) returned in the F2 generation because the recessive alleles from each parent F1 plant combined to result in some F2 offspring that had two recessive alleles.

6. A dominant allele is an allele whose trait is always expressed. The trait of a recessive allele can only be expressed if an organism contains two recessive alleles for a specific trait. A dominant allele suppresses the trait (or expression) of a recessive allele.

7. A sex-linked trait is a non-sexual trait that is carried on a sex chromosome (even though it is a trait that does not have anything to do with a sexual characteristic). Therefore, the gene that codes for the sex-linked trait is found on a sex chromosome even though that gene does not code for a sexual characteristic.

8. False. A sex-linked disease is a genetic disease caused by a defective gene on a sex chromosome. Most of the genetic diseases are caused by defects in genes on the X chromosome.

9. Since sex-linked diseases are caused mainly by defective genes on the X chromosome, in order for a boy to have a sex-linked disease, only one defective chromosome (the X) needs to be inherited. Boys always inherit their X chromosome from their mothers. Since girls have two X chromosomes, they would need to inherit a defective X chromosome from their mother and one from their father, each of which contains the same defective gene.

10. False. Most genetic diseases are transmitted on autosomes.

11. First the gene is isolated, then the bacterial DNA is cut, opening it. This isolated gene DNA is added to the cut bacterial DNA, which allows the isolated gene to insert into the bacterial DNA.

12. Most genetic diseases are caused by defective genes. We can isolate both the normal gene and the abnormal gene. By isolating the normal gene and "giving" it to a person with a genetic disease, that could result in the person being able to start making the normal gene product (protein) instead of the their abnormal gene product. This would, in effect, cure their genetic disease.

CHAPTER 9

1. According to evolutionary thought, life occurred as the result of random processes in the distant past. Over time, one type of organism acquired new traits (or new genes) through the process of mutation. The new traits coded for by the new genes caused one type of organism to evolve (or transform) into another type of organism.

2. According to evolutionary thought, less complex organisms are able to transform into more complex ones through the process of gene-adding (or information-adding) genetic mutations. This process is known as neo-Darwinism.

3. The problem with neo-Darwinism is there is no scientific basis for the presence of information-adding or gene-adding genetic mutations. All known mutations remove information rather than add it.

4. Creationists believe the Bible is the absolutely true word of God. Since it says in the Bible that God created the world and everything in it, creationists believe that as the explanation for the presence of life and all species on earth. Basically, creationists believe God created everything because He said He did.

5. According to evolutionary thought, as one type of organism is acquiring the traits of a new organism (or as one type of organism is in the process of transforming into a new type of organism), there should exist fossil evidence of organisms that have some traits of the species they are transforming from and some traits they are transforming into. However, there are no transitional forms found in the fossil record.

6. When organisms die and fall to the bottom of a body of water, they are rapidly decomposed such that in a short period of time, there is no organism left to undergo the process of slow fossilization.

7. False. Polystrate fossils provide evidence that fossils can form very quickly.

CHAPTER 10

1. A standard classification system helps to improve communication among biologists who speak different languages, avoids confusion created by using common names, and it makes the information gathering, storing and reporting process more accurate.

2. Aristotle.

3. The current taxonomic system was developed by Carolus Linnaeus in the mid 1700's.

4. A binomial name is composed of the genus and species. The genus is written first and is capitalized. The species is written second and is not capitalized. Binomial names are either written in italics or underlined.

5. True.

6. Taxonomy is the science (or study) of organism classification.

7. There are several possible dichotomus keys that could be constructed. This is one of the possibilities:

    1a. Water vehicle – boat.

    1b. Land vehicle – go to 2.

    2a. No engine – go to 3.

    2b. Engine – go to 4.

    3a. Two wheels – bicycle.

    3b. Three wheels – tricycle.

    4a. Two wheels – motorcycle.

    4b. Three or more wheels – semi.

8. False. Eubacteria have cell walls that contain peptidoglycans.

9. True.

10. False. Although "containing DNA" is a criterion that defines life, viruses do not meet multiple other criteria that living things must meet in order to be considered "alive."

11. The only way a virus can replicate (or make more of itself) is to infect a host cell.

CHAPTER 11

1. The three general categories of Protista are: the animal-like protists, or protozoa; the plant-like protists, or algae; and the fungus-like protists, or slime molds.

2. True.

3. Both algae and plants contain chloroplasts, which perform photosynthesis to make their food.

4. Like animals, protozoans need to eat their food in order to live, and so they are both consumers. Also, protozoan cells and animal cells do not have cell walls.

5. Conjugation is a form of sexual reproduction that certain protists display.

6. False. Plankton supplies far more oxygen to the atmosphere as a result of photosynthesis than plants.

7. Fungal cell walls are made from the molecule called chitin.

8. True.

9. Gills form on the undersurface of mushroom caps. Gills are where spores are produced in mushrooms.

10. A stolon is a hyphus that grows along the surface of bread (or on top of the surface a fungus is growing into).

11. True.

12. Mycorrhizae is the term used to describe the symbiotic relationship of a fungus growing in and around plant roots. The fungal component improves the plant's ability to absorb nutrition. The plant component provides food for the fungus to grow.

13. False. Mycorrhizae represent a symbiotic relationship between a fungus and a plant. However, lichen is a symbiotic relationship between a fungus and an alga.

CHAPTER 12

1. A botanist is a scientist who studies plants (or botany).

2. False. Plant cell walls are made of cellulose.

3. When a plant has vascular tissue, that means it has tissues that form into special tubes that carry water, nutrients, minerals, and other materials throughout the plant.

4. Nonvascular plants transport substances through the plant using osmosis and diffusion.

5. Xylem and phloem are specialized plant tissues called vascular tissue. Xylem specifically carries materials and water absorbed in the roots "up" to the rest of the plant. Phloem carries glucose and other materials made in the leaves "down" to the rest of the plant.

6. True.

7. Moss is an example of a nonvascular plant.

8. A spore is a tiny reproductive cell that has the ability to grow into a new plant. Spores are formed for asexual reproduction.

9. Cycads and ginkgoes are both examples of gymnosperms. They are vascular, seed-producing, non-flowering plants.

10. The defining feature of a gymnosperm is the cone.

11. False. Gymnosperms are vascular, seed-producing, non-flowering plants.

12. Monocots have one cotyledon, dicots have two. Monocot leaf veins are parallel, dicots are intersecting/branching. Monocot flower petals are in threes, dicots are in fours or fives. Monocot vascular tissue is scattered randomly, dicot vascular tissue is in bundles in a ring. Monocot roots are fibrous or adventitious and dicot root are in the form of a taproot.

13. True.

14. False. Monocots cannot form tree rings; dicots can.

15. The cuticle layer and closing the leaf stoma both limit water loss.

16. False. That description fits a monocot better.

17. This plant is likely a monocot because the petals are in groups of (or multiples of) three (two groups of three equals six) and monocot petals are usually in multiples of three.

CHAPTER 13

1. Stem cutting and grafting are forms of asexual reproduction, or of vegetative reproduction.

2. Sporophytes produce the reproductive structures called spores.

3. Gametophytes produce male and female gametes, or sperm and eggs.

4. False. ALL plants live either as a sporophyte or gametophyte.

5. False. Fern gametophytes are very small, often no larger than a dime. It is the sporophyte phase we are looking at when we see a "fern."

6. A spore is produced by a sporophyte; therefore, a spore grows into a gametophyte.

7. Gametes are produced by gametophytes. The female gamete is the egg and the male gamete is the sperm. When a sperm fertilizes the egg, the zygote grows into a sporophyte.

8. In a flower, the stamen contains the male reproductive structures, and the pistil contains the female.

9. pollination ⟶ fertilization ⟶ zygote in seed ⟶ zygote grows into sporophyte plant

    male and female spores are produced ⟶ male and female spores grow into male and female gametophyte plants

    pollination ⟵ male and female gametophytes produce pollen

10. Turgor pressure is pressure that develops as the result of water pressure. As water accumulates inside of plant cells, it exerts pressure, called turgor pressure. It is important that a plant maintain the proper turgor pressure so it remains healthy and does not wilt.

11. Material moves through xylem as a result of transpiration.

12. True.

13. Tropisms and nastic movements are different because they are caused by different processes in the plant. Tropisms are caused by auxin and nastic movements are caused by changes in turgor pressure.

14. Photoperiodism is the property of certain plant processes responding to the amount of light they receive.

CHAPTER 14

1. Organisms without a spinal column are called invertebrates.

2. Organisms that have spinal columns are called vertebrates.

3. False. All animals do have an extra-cellular matrix and are multicellular, but no animals have cell walls.

4. Yes, the organism is possibly an animal because all animals store excess energy as glycogen.

5. The defining feature of Porifera (sponges) is that their bodies are filled with pores (holes).

6. A nematocyst is a specialized organelle found within a stinging cell, called a cnidocyte. Nematocysts are found only in organisms classified into the phylum Cnidaria.

7. False. Animals that are polyps are usually sessile.

8. A planarian is a type of free-living flatworm in the class Turbellaria, in the phylum Platyhelminthes.

9. Their specialized head, called the scolex, makes tapeworms effective intestinal parasites.

10. The defining feature of an annelid is the multiple, easily visible body segments.

11. True.

12. True.

13. The rigid exoskeleton made of chitin and the jointed appendages.

14. Malpighian tubules are special arthropod structures that remove wastes from their bodies.

15. During the larval stage, the larva (called a caterpillar) undergoes several moltings (or shedding of the exoskeleton as the larva grows). At the last molting, it takes on the form of a pupa. The pupa then makes a chrysalis, which encloses and protects the pupa. The final metamorphosis occurs within the chrysalis in which the pupa is transformed into a butterfly.

16. False. A nymph is an immature form of an arthropod which undergoes incomplete metamorphosis. Since butterflies undergo complete metamorphosis, nymph would be the improper word for an immature butterfly.

17. Ossicles are the plates that make up an echinoderm's endoskeleton.

CHAPTER 15

1. Cartilage is softer and more flexible than bone.

2. Chordates have an endoskeleton. That means the skeleton is on the inside of the animal.

3. True.

4. Endothermic organisms perform more metabolic reactions than ectothermic ones do. As a result, endotherms produce their own body heat, but ectotherms cannot. Endothermic organisms can regulate their body temperatures independent of the external environment. Ectotherms depend upon the environmental temperature to properly regulate their body temperature. Because of this, endotherms are called "warm-blooded" amd ectotherms are called "cold-blooded".

   The classes Chondrichthyes, Osteichthyes, Amphibia, and Reptilia are all composed of ectothermic organisms. Aves and Mammalia are the only two classes of endothermic organisms.

5. There are a couple of potentially correct answers for this question. First, gas exchange occurs in capillaries. Also, gas exchange takes place in tissues and the lungs/gills.

6. In the lung capillaries, the blood releases carbon dioxide into the air within the lungs. In addition, oxygen is absorbed into the blood from the air in the lungs. Therefore, blood in the lung capillaries becomes oxygenated. The opposite occurs in the tissues. Blood releases oxygen to the tissues and absorbs carbon dioxide from the tissues. Therefore, the blood becomes deoxygenated in the tissues.

7. Veins carry blood to the heart and arteries carry blood away from the heart.

8. Deoxygenated blood is pumped out of the heart. The deoxygenated blood travels to the gills, where it is oxygenated. It continues through the gills to the tissues, where oxygen is released from the blood into the tissues and carbon dioxide is absorbed from the tissues into the blood. The deoxygenated blood continues from the tissues and returns to the heart.

9. True.

10. External fertilization is the type of fertilization that occurs when the eggs are fertilized outside of the female's body.

11. False. Sharks and rays are types of cartilage fish, but seahorses are a species of boney fish.

12. A lateral line is a sensitive collection of nerve tissue that runs along the sides of a fish. It collects sensory information about the environment and sends the information to the fish's brain for processing.

13. False. Amphibian eggs are very sensitive to dryness which is why they need to be laid in or near water.

14. The larval stage of a frog is called a tadpole.

15. The amniotic egg has a tough, protective coating called a shell to protect the embryo. Secondly, the shell has holes in it for gas exchange to occur between the embryo and the environment. Finally, there are other tissues in the egg that remove wastes from and provide nutrition to the embryo.

16. A turtle's shell is made from ribs and covered by tough skin.

17. Alligators have a short, broad snout and crocodiles have a longer and thinner snout. Also, crocodiles have a notch in their upper jaw that the fourth tooth in the lower jaw fits into. Alligators do not have this notch.

18. A snake's tongue is specialized to collect smell (odor) information as it flicks in and out of the snake's mouth. Odor chemicals (molecules) stick to the snake's tongue and are transferred to receptors in the mouth, where they are further processed.

19. False. Crocodiles and alligators have a four-chambered heart. All other reptiles have three-chambered hearts.

CHAPTER 16

1. Birds are placed into the class Aves. Mammals are classified into the class Mammalia.

2. All birds: are endothermic; have feathers, wings, and a beak; lay eggs, have air sacs and a light-weight skeleton. The four unique features are feathers, beaks, air sacs and light-weight skeletons.

3. When a bird breathes in (inhales) fresh air fills the lungs and the air sacs. Gas exchange then occurs in the lungs. When the bird exhales, the old air exits the lungs and the air that was in the air sacs enters the lungs. Gas exchange can then occur again since the air that was in the air sacs still contains oxygen. This allows birds to perform gas exchange during inhalation and exhalation.

4. Webbed feet indicate the bird is a good swimmer (i.e. is an aquatic bird). The webbing of the feet allows the bird to propel itself very powerfully and efficiently with each kick of the foot.

5. Powerful claws are called talons. Birds that have talons are not good swimmers because there is no webbing. Birds that have talons are hunters and are called birds of prey.

6. A keel is a thick breastbone that is specialized for the attachment of the thick flying muscles.

7. True.

8. There are three groups of mammals; the monotremes, the marsupials, and the placental mammals.

9. Monotremes are the egg-laying mammals.

10. Mammals: have mammary glands, hair, specialized teeth, and a single lower jaw bone; are endothermic; store excess energy as fat; have large brains, a diaphragm, and the highest degree of parental care of all animal species.

11. The defining feature of mammals is the presence of mammary glands.

12. The mandible is the lower jaw bone. Mammalian mandibles are special because they are formed from only one bone.

13. False. Both placental mammals and marsupials form a placenta. However, the marsupial placenta does not last very long because the young are born at such an immature stage. The placenta of placental mammals lasts much longer, until the baby is fully developed and ready to be born.

14. The umbilical cord is the connection between the developing mammal baby and the placenta.

15. Rodentia.

16. Insects.

17. A long, sticky tongue.

18. The members of Chiroptera are called bats. They are the only mammals that are able to fly.

19. False. Bats and Cetaceans are able to use echolocation.

20. This is an even-toed ungulate, classified into Artiodactyla.

21. True.

22. Carnivores eat mainly meat. Herbivores eat plants and omnivores eat meat and plants.

23. False. Cetaceans breathe through blowholes but sirens breathe through noses.

24. Primates are the only species to have five fingers per hand/foot covered with a nail.

25. False. No apes have tails.

26. Monkeys usually have tails, apes never do. Ape arms can be lifted all the way over the head for swinging, monkey arms cannot. Ape arms are usually as long/longer than their legs, arms are not. Apes are usually "larger" and monkeys are usually "smaller."

CHAPTER 17

1. The nervous system controls everything that occurs in our bodies.

2. The basic functional unit of the nervous system is the neuron (or nerve cell).

3. See Figure 17.3.1.

4. Dendrites carry impulses (information) to the cell body and axons carry impulses away from the cell body.

5. Sensory nerves make you feel, taste, see, hear, or smell something, motor nerves carry the impulses to make muscles contract, and interneurons are nerves that connect other nerves together.

6. A peripheral nerve is composed of bundles of axons all grouped together to form one larger nerve.

7. The peripheral nervous system is composed of sensory and motor axons bundled together and the central nervous system is composed of the brain and spinal cord.

8. The protective tissues surrounding the CNS are called meninges.

9. The human cerebrum is larger than the cerebellum.

10. The temporal lobes are the areas of the brain concerned with hearing.

11. The occipital lobes process vision.

12. Grey matter is located in the periphery of the brain, or the outer layer of brain tissue. It looks grey because it contains cell bodies and not myelin.

13. A reflex arc is composed of a sensory nerve, an interneuron, and a motor nerve.

14. Sound enters the ear through the auditory canal and hits the tympanic membrane (ear drum), causing it to vibrate. As the tympanic membrane vibrates, the middle ear bones connected to the tympanic membrane also vibrate. The middle ear bones are also connected to the cochlea. As the middle ear bones vibrate, the cochlea does, too. The vibration within the cochlea causes the cochlear fluid to move, which moves cilia that line the interior of the cochlea. The movement of the cilia causes nerves to send impulses to the brain, which are then processed as sound.

15. False. Rods are responsible for sensing light. Cones are responsible for processing color.

16. See Figure 17.12.1.

17. Cornea, pupil, lens, vitreous, retina.

18. A target tissue is the tissue a hormone acts upon.

19. A receptor is a molecule on the surface of a target tissue to which a hormone binds.

20. True.

21. The dermis is lower than the epidermis.

22. True.

23. Keratin is the tough protein that makes up a large part of epithelial (skin) cells.

24. The three muscle cell types are cardiac muscle, skeletal muscle, and smooth muscle.

25. A tendon connects a muscle to a bone.

26. True.

27. The radius is a bone in the arm (the forearm).

28. Bone marrow is located in the hollow cavity of bone. It makes blood cells; specifically, red blood cells and white blood cells.

29. Blood vessels pass through bones via small canals in the bone.

30. An osteocyte is a bone cell.

31. A ligament is a type of connective tissue that holds one bone to another where the bones meet to form a joint.

32. A ball and socket joint has more movement than a hinge joint.

CHAPTER 18

1. Vitamin D can be found in high concentration in fish oil, liver, milk, and eggs. Vitamin B12 is found in green vegetables and liver.

2. Mechanical digestion is the physical breakdown of food. Chemical digestion is the enzymatic breakdown of food, combined with the acidity of the stomach. The teeth and stomach are the main contributors to mechanical digestion. The salivary glands, stomach, small intestine, liver, gall bladder, and pancreas contribute to the chemical digestion process.

3. False. The spleen is part of the immune system.

4. The gall bladder stores bile (made in the liver). When we eat, the gall bladder contracts and squeezes the bile into the small intestine.

5. Peristalsis refers to the waves of muscular contractions that occur throughout the GI system. Peristalsis is responsible for propelling the food all the way from the esophagus through the large intestine.

6. Villi and microvilli are folds in the small intestine lining. Villi are the larger folds and microvilli are the smaller folds that form on the villi. They are responsible for absorbing the nutrients we eat.

7. The blood starts in the right atrium and moves through the tricuspid valve, into the right ventricle, through the pulmonic valve to the pulmonary arteries, into the lungs, into the pulmonary veins, into the left atrium, through the mitral valve, into the left ventricle, through the aortic valve into the aorta and arteries and into the systemic capillaries. Then it moves into the veins and returns to the right atrium through the superior or inferior vena cavae.

8. All valves are one-way. The valve is supposed to open to allow blood to flow through and then close to prevent the back-flow of blood.

9. False. The pulmonic valve is between the right ventricle and the pulmonary artery.

10. False. The mitral valve needs to be open when the left atrium pumps to allow the blood to flow from the left atrium into the left ventricle. The mitral valve is located between the left atrium and left ventricle. If the mitral valve is closed when the left atrium contracts, the blood could not flow from the atrium into the ventricle.

11. The sinoatrial node, or SA node, is the pacemaker of the heart.

12. Plasma, white blood cells and red blood cells.

13. The special property of hemoglobin is its ability to bind to oxygen and carry it from the lungs to the tissues. Hemoglobin is found within red blood cells.

14. An alveolus is the functional unit of the lung.

15. During inhalation, the diaphragm contracts, which causes it to move downwards. As it moves down, it increases the chest-space size and creates a negative pressure (like a vacuum) within the chest cavity. This causes the lungs to expand to fill the chest cage. As they expand, a negative pressure is created within the lungs which sucks air into them. At the end of inhalation, the lungs are stretched tight because they have expanded with air. With the relaxation of the diaphragm, the stretched-tight lungs naturally relax and deflate, forcing the air out and causing exhalation.

16. The epiglottis is a fold of skin that protects the airway from food and liquid that are being swallowed so it doesn't go into the larynx and trachea. During a swallow, the muscles pull the larynx (voice box) upward, which closes the airway between the epiglottis and the top of the larynx.

17. False. When the diaphragm contracts, we inhale. When the diaphragm relaxes we exhale.

18. The excretory system is composed of the kidneys, ureters, bladder, and urethra.

19. A nephron is the functional unit of the kidney.

20. No. Recall waste removal occurs in two basic steps. The first step is a blood-filtering step in which all wastes, water, ions, and nutrients are filtered out and removed from the blood. The second step is a reabsorption process in which ions, nutrients, and water are reabsorbed from the waste area back into the blood. This removes these valuable materials from the urine. Therefore, if the nephron filters these substances out but does not reabsorb them, it is not working the way it is supposed to.

21. True.

22. During the active phase of the immune response, the immune system is specifically targeting pathogens for removal. Inflammation occurs, releasing chemicals that serve as calling signals for white blood cells to infiltrate the area. When the white blood cells enter the infected area, they actively kill the pathogens. Also during the active phase, special proteins called antibodies are released into our bodies. Antibodies help white blood cells target specific pathogens for removal.

23. An autoimmune disease is a disease in which a person's own immune system is stimulated to attack tissues in their own body.

CHAPTER 19

1. The biosphere is the area of earth that can support life.

2. Ecology is the study of how living organisms interact with one another and their environment.

3. An ecosystem is the association and interaction of all living organisms within their physical environment. The boundaries of an ecosystem (which is the physical area of the ecosystem) are defined by the scientist who is studying the ecosystem.

4. The two general components of an ecosystem are the biotic and abiotic mass (or biotic and abiotic components).

5. Rock, sand, and water are all examples of abiotic mass.

6. See Figure 19.5.2.

7. True.

8. False. Burning decreases the amount of oxygen in the air. It increases the amount of carbon dioxide in the air.

9. Biodegradeable pollutants can be broken down by organisms that live in the ecosystem; non-biodegradeable pollutants cannot.

10. A community includes more organisms than a population (i.e. communities are larger than populations).

11. A food web is a fairly accurate depiction or description of the energy transfer that occurs between producer, consumer, and decomposer in an ecosystem. Basically, a food web shows "who eats who" in an ecosystem.

12. True.

13. Mimicry is when a non-harmful organism resembles a harmful one or when a harmful organism resembles a non-harmful organism. Camoflage is when an organism is shaped and/or colored to blend into its environment.

14. There are seven established terrestrial biomes. However, do not forget the aquatic biomes. There are many different marine and fresh-water biomes.

15. The temperate grassland biome.

16. Permafrost is a layer of soil that is permanently frozen. It is found only in tundra.

17. In order for succession to occur, an ecosystem needs to be destroyed or needs to be a potential ecosystem with no species living in it. Initially, brand new species begin to inhabit the ecosystem. These are the pioneer species. Over time, more and more species move into the ecosystem. This is termed primary succession. As more time passes, still more species move into the system and begin to replace those there before. This is called secondary succession. Over time, there will be no net change in the species that inhabit the ecosystem and when this happens, the climax community is reached.

Life Science Test Answer Key

Note to parent:

These tests are purposely designed to be in-depth knowledge assessment tools. As such, they are fairly long, so please do not be intimidated! Essay or short answer-type questions are the best way to identify whether or not your student completely understands the material. It is perfectly acceptable, if you choose, to have your student answer only the even questions, only the odd questions, or every third question as you see fit for your student's need. If they are able to correctly answer the majority of questions asked, it is likely they understand all of the material very well. If your student answers questions better orally, please ask them that way, rather than having them write it out. The method of test taking should not be a stumbling block in your student's performance.

LIFE SCIENCE TEST #1 (Chapters 1-3)

1. What do biologists study? **Living organisms.**

2. What is the process of the scientific method? **A scientist or student develops a question regarding observations made about anything. The question is formulated into a testable hypothesis, then experimentation begins to test the hypothesis. As data is collected during experimentation, it is analyzed to test whether the data proves the hypothesis True or False. If the data supports the hypothesis, further experiments are designed and carried out to test the hypothesis more. If the hypothesis is not supported by the data, the hypothesis must either be re-formulated or thrown out altogether, and a new hypothesis statement made. If a hypothesis holds up over many years and experiments, it is considered a theory.**

3. True or False? The experimental group is exposed to the variable in a controlled experiment. **True.**

4. What is the difference between a measured and a derived unit? **Measured units can be measured directly using some measuring device; derived units must be calculated because they represent mathematical relationships between two or more measured or other derived units.**

5. List at least five ways data can be obtained during an experiment. **Light microscopy, electron microscopy, thermometer, scales, speedometers, cameras, unaided vision, hearing, taste, touch, smell, CT scanning, MRI, and x ray (to name a few).**

6. True or False? In order to study life, one must know what is alive and what is not. **True. Establishing common criteria to define life allows for accurate study and discussion of the experimentation process.**

7. True or False? A multicellular organism is made up of two or more cells. **True.**

8. True or False? Unicellular organisms are not alive because they only contain one cell. **False. Unicellular organisms are every bit as alive as multicellular organisms.**

9. Why is DNA called the "blueprint of life?" **DNA contains the genetic information that codes for, or controls, the development of every aspect of an organism's appearance and function.**

10. What is the difference between sexual and asexual reproduction? **Sexual reproduction occurs by the DNA of two organisms combining to form a new organism. Asexual reproduction occurs without combining genetic information. Sexual reproduction forms organisms that are genetically different than either parent, but asexual reproduction leads to organisms that are genetically identical to the parent cell.**

11. True or False? Even unicellular organisms are complex and organized. **True. All life forms are complex and organized no matter how big or small they are.**

12. Do multicellular organisms have a higher degree of complexity than single cell organisms? **Of course. They are larger and have more cells, so they are naturally more complex and organized relative to unicellular organisms.**

13. What are the structures called that allow an organism to sense and respond to their environment? **Receptors.**

14. True or False? Plants convert the sun's energy through the process of cellular respiration. **False. Plants convert the sun's energy through photosynthesis.**

15. True or False? Plants extract the energy they need from their environment, but animals do not. **False. All living things extract the energy they need to live from their environment.**

16. When a fish that is cold swims into warmer water to warm up, then later swims to cooler water to cool down, what property of life is the fish displaying? **Homeostasis.**

17. What is a binomial name? **It is a specific name for an organism that no other organism shares. The binomial name is a unique identifier. The first part of a binomial name is the genus the organism is classified into; the second name is the species.**

18. What is an aerobic organism? **One that needs oxygen to survive.**

19. True or False? Air is made up mainly of nitrogen, oxygen, carbon dioxide, and argon. **True.**

20. Why are matter, mass, and weight not the same thing? **Matter is anything that takes up space. It is the physical presence of atoms and molecules. Mass is the amount of matter that something has. Unless part of the object is broken off of the original, mass does not change (under "normal" conditions). Weight is dependent on the pull of gravity that is acting on the object that contains matter and has mass. Weight changes depending on the gravitational pull, but matter and mass do not.**

21. What is one way that you can prove something you cannot see, such as a gas, contains matter? **Blow up a balloon. If the gas you were blowing into the balloon did not have matter, then the balloon would not fill. The matter in the gas takes up space inside the balloon, and the more matter you blow into the balloon, the fuller it gets.**

22. True or False? A molecule is matter composed of only one type of atom. **False. A molecule is made up of two or more atoms linked together. An element is matter composed of only one type of atom.**

23. What is an organic molecule? **Molecules that contain carbon and are normally made by living cells.**

24. What are the four types of organic molecules? **Carbohydrates, lipids, proteins, and nucleic acids.**

25. What three atoms compose all carbohydrate molecules? **Carbon, hydrogen, and oxygen.**

26. True or False? Glucose is an example of a complex carbohydrate. **False. Glucose is a simple carbohydrate.**

27. True or False? Lipids are usually solid at room temperature, while fats are usually liquid. **False. Fats are usually solid and oils are usually liquid.**

28. How are proteins made? **By linking together smaller units called amino acids.**

29. True or False? Proteins are not sensitive to changes in their amino acids, so the amino acids can be changed in a protein without much effect, if any, to the organism or the way the protein works. **False. Proteins are sensitive to small changes in their amino acids, so much so that sometimes just changing one amino acid can make a protein or enzyme completely non-functional.**

30. Why is it said that the sun is the ultimate source of all biological energy on earth? **Without the sun, plants would not perform photosynthesis, so they would not make any glucose molecules. Without glucose molecules, neither producers nor consumers would have anything to use during cellular respiration. Without cellular respiration, no organisms would be able to make ATP, and without ATP, no chemical reactions would happen. Life would end on earth without the sun.**

31. True or False? Plants use the energy they reflect for photosynthesis. **False. Plants can only use the energy they absorb. Reflected light energy is lost light energy.**

32. True or False? Objects appear the color they are because of the light they reflect. **True.**

LIFE SCIENCE TEST #2 (Chapters 4-5)

1. What is the outer boundary of the cell called? **The cell membrane.**

2. True or False? Cell theory states there is no smaller unit capable of life than a cell. **True.**

3. True or False? Cells are made up of non-living molecules. **True.**

4. Why are cells described as spheres or cubes rather than as circles and squares? **Because cells are three-dimensional structures and should be described with three-dimensional words such as** *cube* **and** *sphere*, **rather than two-dimensional words such as** *circle* **and** *square*.

5. What do you call a group of cells working together with a similar function? **A tissue.**

6. What is an organ system? **Two or more organs working together to accomplish a similar function for the good of the organism.**

7. What is a prokaryote? **An organism that is composed of a prokaryotic cell.**

8. How does a prokaryotic cell differ from a eukaryotic cell? **A prokaryote does not have a nucleus, nor does it have membrane-bound organelles. Eukaryotes have both.**

9. What is a nucleoid? **The area of the cytoplasm where prokaryotic DNA is contained.**

10. Why do phospholipids orient with their tails facing toward and their heads facing away from one another? **All cells are surrounded by a watery environment. The phospholipid heads like water, and the tails do not. When many phospholipid molecules are placed into a watery environment, the tails naturally align toward the center of the lipid bilayer to get as far from the water as they can. The heads do not mind at all sitting out in the water, so they face outward, toward the water.**

11. What is meant by "the membrane is selectively permeable?" **It means only certain molecules can get across—the membrane is picky and does not let just any molecule cross into or out of the cell. There are some molecules able to freely pass through the membrane, but most cannot.**

12. True or False? Diffusion is the random movement of molecules from an area of high concentration to an area of low concentration. **True.**

13. What is meant when it is said that equilibrium has been established relative to diffusion. **It means the concentration of molecules is equal on each side of the membrane.**

14. What is osmosis? **It is the diffusion of water across a selectively permeable membrane from an area of high water concentration to low concentration.**

15. What is the difference between active and passive transport? **Passive transport occurs without the cell using any energy. Active transport requires the cell to use energy.**

16. True or False? Cell pump proteins actively move molecules from areas of high concentration to areas of low concentration. **True.**

17. True or False? Endocytosis expels items from the cell and exocytosis brings them into the cell. **False. Endocytosis brings molecules into the cell and exocytosis expels them from the cell.**

18. What is an organelle? **It is the individual functional unit of the cell. Every type of organelle has a specific function in the cell.**

19. If you were looking through a microscope at a cell with a nucleus, what type of cell would it be: prokaryotic or eukaryotic? **Eukaryotic.**

20. What are two differences between a plant and animal cell? **Plant cells have cell walls and chloroplasts, animal cells do not.**

21. Why is it important that the cytoplasm is somewhat watery? **The fluid nature of the cytoplasm allows for wastes, molecules, and organelles to move around in the cytoplasm.**

22. What is the nucleolus and what does it do? **It is an area in the nucleus that manufactures pieces of ribosomes.**

23. What are nuclear pores? **They are small openings in the nuclear membrane that allow molecules to pass into and out of the nucleus.**

24. Do ribosomes have a membrane? **No, they are the only non-membrane-bound organelle in a eukaryotic cell.**

25. What do ribosomes do? **They make proteins using instructions contained in mRNA.**

26. True or False? ER packages and transports proteins after they are made by ribosomes. **True.**

27. What is the name for the series of flattened and stacked tubes with small sacs at the end of them? **The Golgi apparatus.**

28. Where are lysosomes made? **In the Golgi apparatus.**

29. What are the inner folds of membranes called in mitochondria? **Cristae.**

30. What happens in the cristae? **They are where ATP is made.**

31. What is the function of a vacuole? **It is for storage of substances in the cell.**

32. What important photosynthetic molecule do chloroplasts contain? **Chlorophyll.**

33. True or False? ATP is the molecule made during photosynthesis. **False. ATP production is a function of mitochondria.**

34. What is important about the glucose plants make during photosynthesis? **The glucose provides all the available glucose for cellular respiration to occur in animals and other consumers.**

35. True or False? During photosynthesis, six molecules of water and six molecules of carbon dioxide are used to make one molecule of glucose and six molecules of oxygen. **True.**

36. What generally occurs during cellular respiration? **Glucose is broken down by mitochondria. This releases energy, which is captured and used to make many molecules of ATP.**

37. What is important about ATP? **It is the molecule that almost every cell on earth uses for energy.**

38. What is the relationship between cellular respiration and photosynthesis? **Cellular respiration releases carbon dioxide into the air, which plants need to use during photosynthesis. Photosynthesis releases oxygen into the air, which is needed during cellular respiration. Also, photosynthesis produces glucose, which is broken down during cellular respiration to make ATP.**

39. True or False? When oxygen supply is low, cells begin to make ATP through fermentation. **True.**

40. True or False? Fermentation makes more ATP per molecule of glucose than aerobic cellular respiration. **False. Fermentation makes it quicker, but much less of it per molecule of glucose than aerobic respiration.**

LIFE SCIENCE TEST #3 (Chapters 6-7)

1. What are the four nucleotides of DNA? **Adenine, guanine, thymine, and cytosine.**

2. What are cells that contain two pairs of each chromosome called? **Diploid cells.**

3. Why are chromosomes paired? **Because each chromosome of a pair contains genes for the same trait.**

4. How many chromosomes, total, do humans have? **Forty-six.**

5. How many pairs of chromosomes do humans have? **Twenty-three.**

6. If a person has an X and a Y chromosome, is that person a male or female? **Males are XY and females are XX.**

7. True or False? Chromosomes are the smallest unit of the genetic code. **False. Genes are the smallest unit of the code and are contained in chromosomes.**

8. True or False? Genes are segments of a chromosome that contained on opposite strands of DNA so one gene is contained on both strands of DNA. **False. A gene is on one DNA strand only. One gene is not partly contained on one strand with the rest on another strand.**

9. How many nucleotides are contained in a codon? **Three.**

10. True or False? A gene is a segment of DNA that contains codons. **True.**

11. Where is mRNA made? **The nucleus.**

12. Where are proteins made? **The cytoplasm.**

13. What cell organelle makes proteins? **Ribosomes.**

14. How do ribosomes know how to make the right protein? **They decode the message in the mRNA that tells them which amino acids to link together. Ribosomes simply follow the instructions they read in the codons of the mRNA.**

15. What nucleotide base pairs with adenine in mRNA? **Uracil.**

16. How is a molecule of mRNA made? **Special proteins, called enzymes, unzip the DNA that needs to be made into a molecule of mRNA. This is the gene that codes for the protein that is to be made. Other enzymes make mRNA from DNA—exactly transferring the codon sequence from DNA to mRNA to preserve the genetic code.**

17. True or False? Ribosomes "read" the mRNA molecule as we would read a book—one way, codon by codon (letter by letter in the case of reading a book). **True.**

18. What is another word for protein synthesis? **Translation.**

19. Where in the cell does translation occur? **The cytoplasm.**

20. True or False? tRNA is a molecule that shuttles amino acids to the ribosomes. **True.**

21. True or False? When DNA is replicated, one strand is copied first, then the opposite strand is copied. **False. Both strands are copied at the same time.**

22. Why would a cell need to copy its DNA? **So when one cell divides into two cells, each of the two cells receives a copy of the instructions telling them how to function properly.**

23. Mutations are not necessarily bad for an organism. Why? **The mutation may not ever be noticed if it occurs in an area of the DNA that does not contain a gene.**

24. True or False? There are a number of diseases that are caused by mutations. **True.**

25. Which type of mutation, point or chromosomal, involves more DNA? **Chromosomal.**

26. What is the general process of cell division? **One parent cell divides into two daughter cells.**

27. What is the general process of cell reproduction, including what occurs to the DNA? **The DNA in the parent cell is replicated so the cell contains two exact copies of DNA. The DNA is then separated to opposite ends of the cell so one end receives a complete copy. Then the parent cell divides into daughter cells so that each of the daughter cells contains a complete copy of DNA (chromosomes).**

28. What is it called when the chromosomes (DNA) separate during cell division? **Karyokinesis.**

29. Describe the process of binary fission. **The prokaryote cell copies its DNA, then begins to elongate. As the cell gets longer, the two copies of DNA move farther away from one another. The cell then divides into two cells so that each daughter cell contains an exact copy of the DNA.**

30. True or False? When a multicellular organism buds, a mini version of the adult grows from the adult organism, then falls off. **True.**

31. True or False? The stages of the cell cycle are divided based on what is happening to the cell membrane during mitosis. **False. The cell cycle is divided into stages based on what is happening to the chromosomes.**

32. When do the chromosomes get pulled apart during mitosis? **Anaphase.**

33. True of false? Haploid cells contain one copy of every chromosome in them and diploid cells contain two copies of every chromosome. **True.**

34. True or False? Gametes are diploid and body cells are haploid. **False. Body cells are diploid and gametes are haploid.**

35. Why do gametes have to be haploid cells. **So that when the chromosomes combine during fertilization, the zygote has the proper number of chromosomes. If gametes were diploid, the zygote would have twice as many chromosomes as is normal.**

36. True or False? Meiosis results in the production of four haploid gamete cells from one diploid parent reproductive cell. **True.**

LIFE SCIENCE TEST #4 (Chapter 8-9)

1. What is a trait? **A characteristic of an organism.**

2. True or False? Geneticists study the passage of traits from generation to generation. **True.**

3. What is the P generation? **It is the starting generation of a genetic study.**

4. True or False? A hybrid is an organism that produces offspring with exactly the same traits over time. **False. A hybrid is an organism that contains genes from two different purebred parents. This means they have genes that code for different traits and will produce offspring with different traits over many generations. A purebred is an organism that produces offspring with the same traits generation after generation.**

5. True or False? The yellow pea trait is recessive to the green pea trait. **False. The yellow pea trait is the dominant pea color.**

6. When a dominant and recessive allele are present in the same organism, which trait is expressed and which trait is suppressed? **The dominant trait is expressed and the recessive trait is suppressed.**

7. If an organism has the following two alleles for a characteristic—GG—is this organism a purebred or hybrid? **This is a purebred because the alleles are both the same.**

8. Why did the recessive traits Mendel was studying disappear in the F1 generation? **Because all the offspring were hybrids, meaning each organism in the F1 generation had one recessive and one dominant allele. There were no other combinations of alleles in the F1 generation. No matter which trait Mendel studied, the F1 generation always had one recessive and one dominant allele for the condition. The dominant allele "dominated" the recessive, so none of the F1 organisms displayed any recessive traits.**

9. True or False? In order to use a Punnett square, the allele types of both parents must be known. **True. The allele type of one parent is written at the top of the square and the allele type of the other parent is written on the side of the square. The information is then filled in the boxes.**

10. What is the reason that Mendel found the recessive trait reemerged in the F2 organisms? **Since each F1 organism contained a recessive allele, during gamete formation, there were gametes that formed containing only the recessive allele for the trait. If a male parent recessive gamete and a female parent recessive gamete from the F1 generation which each contained a recessive allele combined with one another, then that F2 offspring would have both recessive alleles for the condition and would display the recessive trait.**

11. What happens when a red flower gamete combines with a white flower gamete, and the alleles have an incomplete dominance-type of relationship? **The result is a blend of the traits of the two alleles. For example, when a red flower and a white flower are crossed, the result is a pink flower. The pink is a blend of the red and the white.**

12. What is the term for the condition in which more than one gene affects the way a trait is expressed? **Multiple gene inheritance pattern.**

13. True or False? A sex-linked trait is carried on a sex chromosome, but does not control the development of a sexual characteristic. **True.**

14. Why are males more likely to have a sex-linked disease than females? **The sex-linked diseases are almost always carried on the X chromosome. Females have two X chromosomes, and it is unlikely that a female will inherit a defective sex-linked trait from her mother and her father. If she only has one sex-linked gene that is defective, the normal sex-linked gene on the other X chromosome will counteract it. Males only have one X chromosome, always inherited from the mother. If that one X chromosome has a defective sex-linked gene, there is no normal gene to counteract it on the Y chromosome, so the male is much more likely to have a sex-linked genetic disease.**

15. True or False? A female is considered a carrier if she has a defective sex-linked gene but does not have the disease for which it codes. **True.**

16. True or False? Duchenne muscular dystrophy is caused by a sex-linked gene that codes for the production of a defective muscle protein. **True.**

17. How many defective alleles does a person need to have in order to have an autosomal genetic disease? **Two—both alleles that code for the same protein need to be defective.**

18. True or False? Genes are altered and made abnormal due to infections. **False. Mutations cause defective genes.**

19. Down syndrome is a type of what genetic disease? **A chromosomal genetic disease. A person with Down syndrome has three chromosome number 21 instead of two.**

20. True or False? Selective breeding is an example of genetic engineering. **True.**

21. Can an organism be given a trait it does not have? **Yes, that is the basis of genetic engineering—figuring out a way to insert a gene into something and give it a trait it does not have.**

22. True or False? Evolutionists believe that life is on this planet for a reason and creationists believe life exists here by events that occurred through random chance. **False. Evolution is the belief that life arose on the planet through random events. Creation is the belief that life originated through the specific creation by God and that all life has a purpose.**

23. True or False? Evolution is an atheist's way of describing how life originated on earth. **True.**

24. True or False? Charles Darwin is considered the father of evolution. **True.**

25. True or False? Evolutionists believe that the first life-form was single-celled structure, such as a prokaryote, which assembled into a cell randomly. **True.**

26. Evolutionists believe the earth is billions of years old. **True.**

27. Can organisms evolve if they do not acquire the genes that code for the new traits that cause them to evolve? **No. In order for evolution to cause a less-complicated organism to evolve into a more complicated one, the less-complicated organism needs to acquire new genes that code for the development of the more complicated species' structures and actions.**

28. How do evolutionists propose that less-complicated organisms acquire new genes? **Through mutations.**

29. True or False? According to the phylogenetic tree, humans are most closely related to gorillas. **False. According to the phylogenetic tree, humans are most closely related to chimpanzees.**

30. Evolution is based on hard fact and the basic belief of neo-Darwinism has been proven over and over again. **False. Evolution relies on faith that certain things happened the way evolutionists believe they did just as creation is based on faith that certain things happened the way creationists believe they did.**

31. What does intelligent design mean? **Since God is all-powerful, all-knowing, and knows what the plans are for everything He creates, He would have created every organism with the exact amount of DNA and genes it needs to fulfill His plan.**

32. What objections do evolutionists have to creation? **Evolutionists do not believe it is scientific to believe in something that cannot be proven, and they do not believe the existence of God can be proven. Therefore to believe in His creation is not scientific.**

33. True or False? The belief that information adding mutations are an explanation for the diversity of life is based on scientific fact. **False. There has never been one information-adding mutation proven to occur, nor has there ever been one induced in a scientific setting. Evolutionist's have a faith-based belief that information-adding mutations occur since they have never been proven to occur.**

34. True or False? Creationists do not believe that natural selection occurs. **False. They do believe it occurs.**

35. Is the example of a dark-living species of animal losing their eyes over time evolution? Why or why not? **It is not evolution. Evolution is by definition a less complex organism being transformed into a more complex organism by gene-adding mutations. When the dark-living organisms lost their eyes, they did so through mutation. However, the mutations caused two things: a loss of genetic information (the information coding for the development of eyes), and the movement of a more complex organism into a less complex one (an organism without eyes is obviously less complex than one with eyes). This is not evolution.**

36. True or False? If information-adding mutations did occur, they would likely make an organism less likely to survive in their environment and cause them to be killed through natural selection. **True.**

37. True or False? A transitional form is an animal that shows traits of the organism it is evolving from and changing into. **True.**

38. True or False? The fossil record is full of transitional forms. **False. There is not one fossil that has been accepted to be a transitional form.**

39. True or False? The reason that evolution is so appealing as an explanation for the origin of life and species is because it is supported by hard scientific facts. **False. Evolution requires as much faith to believe as creation.**

LIFE SCIENCE TEST #5 (Chapters 10-11)

1. True or False? A kingdom is the largest level of categorization. **True.**

2. Why is a binomial name helpful in the seven-level system? **Since there is only one species that has a given binomial name, there is no confusion when scientists describe the organisms they are studying.**

3. Are dichotomous keys helpful for a person to identify types of bacteria as well as animals? **Yes. They are helpful to identify any type of living organism.**

4. What are the three cell shapes that bacteria display? **Spiral, coccus (spherical or round) and bacillus (rod).**

5. What are peptidoglycans? **They are the molecules that make up the bacterial cell wall.**

6. What is a virus considered if it is not a living organism? **A particle.**

7. What is a host and why does a virus need one? **A host is the cell or organism a virus infects. Viruses need hosts so they can take over the protein-making machinery of the host cell and make more of the virus.**

8. What are three types of protists? **Animal-like protists (protozoa), plant-like protists (algae), and fungus-like protists (slime molds).**

9. True or False? Amoeba and paramecium are common types of protozoans. **True.**

10. True or False? The plant-like protists, algae, do not perform photosynthesis. **False. Algae contain chloroplasts and chlorophyll and perform photosynthesis.**

11. Why are slime molds called fungus-like protists? **They both obtain their energy the same way, through the digestion of dead material (they are both decomposers).**

12. What is the difference between conjugation and fragmentation? **They are performed by different protist species. Fragmentation is performed by algae and conjugation by protozoa. Also, fragmentation is asexual reproduction and conjugation is sexual reproduction.**

13. What is a saprophyte? **It is an organism that obtains its nutrition by eating dead organisms. Decomposers are saprophytes.**

14. What is a hyphus? **It is the basic structural unit of fungi and is composed of many hyphal cells linked to one another end to end.**

15. True or False? Fungi can be both helpful and harmful to man. **True.**

16. True or False? The mycelia of mushrooms that sit on top of their food secrete enzymes to digest the food, then the mycelia absorb the released nutrition. **True.**

17. True or False? Stolons are hyphae that grow down into the food source. **False. Stolons are hyphae that extend across the surface of the food source.**

18. True or False? Asexual reproduction of Fungi occurs when hyphae from different fungi of the same species fuse and form a new organism. **False. That is sexual reproduction in Fungi. Asexual reproduction is through spores.**

19. Lichens are a symbiotic relationship of Fungi with what other organism? **Algae.**

20. True or False? Fungi can form symbiotic relationships with plants and algae. **True.**

LIFE SCIENCE TEST #6 (Chapters 12-13)

1. True or False? There are four general categories of plants. **True.**

2. Are the majority of plants vascular or nonvascular? **Vascular.**

3. How do nonvascular plants transport substances through the plant? **Osmosis and diffusion.**

4. True or False? The sexually-reproducing, nonvascular moss plant grows out of the non-sexually-reproducing moss plant. **True.**

5. True or False? Xylem and phloem are continuous tubes found only in the leaves and roots. **False. Vascular tissue runs throughout the plant.**

6. True or False? There are two basic types of vascular plants, those that form seeds and those that do not. **True.**

7. Ferns and horsetails reproduce with spores. **True.**

8. True or False? Ferns only have one form of the plant, in which the plant reproduces only asexually with spores. **False. Ferns have two plant cycles (or plant forms)—a cycle (or form) in which the fern plant reproduces sexually and another cycle (or form) in which the fern plant reproduces asexually with spores.**

9. What is endosperm? **It is the supply of food contained in a seed.**

10. What is a plant that has xylem and phloem and produces seeds but not flowers called? **A gymnosperm (conifer and pine tree are okay answers, too).**

11. True or False? A cone is the asexual reproductive structure of a gymnosperm. **False. It is the sexual reproductive structure.**

12. True or False? Angiosperms are the second most numerous plant species, behind the gymnosperms. **False. They are the most numerous.**

13. What is the difference between a monocot and a dicot? **A monocot plant produces seeds with one cotyledon and a dicot produces seeds with two cotyledons.**

14. What is the difference between monocots and dicots as far as root structure is concerned? **Monocots usually have a fibrous root system and dicots usually have a taproot system.**

15. What is the difference between monocot and dicot stems as far as their vascular bundles (xylem and phloem) are concerned? **Monocot stems have their xylem and phloem scattered throughout the stem. The xylem and phloem in dicots is arranged in a ring around the center of the stem.**

16. True or False? Tree rings form because the tree grows faster in the late summer and fall than in the spring and early summer, which causes the xylem to form and look different. **False. They form because the tree grows faster in the spring and early summer than in the late summer and fall.**

17. What is the difference between the apical and lateral meristems? **Apical meristems are found at the tips of stems and roots. When apical meristem growth occurs, the roots and stems get longer. Meristems are also found between the xylem and phloem. This is called lateral meristem tissue. Growth of the lateral meristem tissue causes plant parts to get larger in girth ("bigger around"). Lateral meristem tissue growth also forms new xylem and phloem.**

18. What is the difference between monocot and dicot leaves regarding the vascular formation? **Monocot vasculature runs parallel, dicot vasculature intersects in a branching pattern.**

19. True or False? Grafting, stolons, and stem cutting are all ways that plants can asexually reproduce. **True.**

20. Which form, or phase, of a plant are gametes produced? **The gametophyte phase.**

21. How is a nonvascular plant sporophyte formed? **The gametophyte plant produces sperm and eggs. A sperm fertilizes an egg and forms a zygote. The zygote then grows into the sporophyte plant.**

22. True or False? The gametophyte form of the nonvascular plants usually grows from the top of the sporophyte form of the plant. **True.**

23. True or False? The large fern plant that is recognizable as a fern is a fern in the sporophyte phase. **True.**

24. Where do fern spores form? **On the undersurface of the leaves (fronds).**

25. True or False? A conifer (gymnosperm) male gamete fertilizes the female gamete in the pine cone. **True.**

26. True or False? A conifer spore grows into a gametophyte. **True.**

27. True or False? Male and female conifer (gymnosperm) gametophytes are large structures. **False. They are small structures that grow in pine cones.**

28. True or False? Plants that are pollinated by the wind often contain nectar. **False. Plants that are pollinated by the wind usually do not have nectar. Plants that are pollinated by insects/animals usually have nectar.**

29. What is the female reproductive structure of a flower called? **The pistil.**

30. What is the male reproductive part of a flower called? **The stamen.**

31. How does fertilization occur in angiosperms (following pollination). **A tube quickly forms from the pollen grain. The tube grows from the pollen through the style to the ovary. A sperm cell moves through this passage to the egg in the ovule. The sperm unites with the egg and the chromosomes from the male parent are combined with those of the female parent. A zygote is formed and is covered by a seed coat.**

32. What is germination? **The process of a seed sprouting and beginning to grow.**

33. Describe the function of guard cells. **The guard cells open and close the stoma on the undersurface of the leaf depending on water conditions. If water is in short supply, the guard cells close the stoma so photosynthesis cannot occur. This saves water so the plant does not dry out. When water is plentiful, the guard cells open the stoma again. By doing so, carbon dioxide can enter the leaf so photosynthesis can occur.**

34. What lost property of a plant is responsible for wilting? **Turgor pressure.**

35. How do nastic movements occur? **Due to changes in turgor pressure.**

36. True or False? Transpiration provides a "pulling" force on water inside of xylem and serves to "pull" water upward in the xylem. **True.**

37. What do plant hormones usually regulate? **Plant cell growth.**

38. What does ethylene do to a fruit? **It causes it to ripen quickly.**

39. How are tropism and nastic movements different? **Tropisms occur due to the effects of hormones; nastic movements occur due to the effects of turgor pressure.**

40. True or False? The time of year that many plants produce flowers is dependent on the amount of light they receive, which is an example photoperiodism. **True.**

LIFE SCIENCE TEST #7 (Chapters 14-16)

1. How do zoologists classify animals? **Zoologists use similarities and differences in an animal's structure and function to classify them.**

2. True or False? Animals have the following properties in common: no cell walls, extracellular matrix, nucleoid, and sexual reproduction. **False. The nucleoid is not an animal feature.**

3. What is the difference between a vertebrate and invertebrate? **A vertebrate has a spinal column, composed of bones called vertebrae, that protects a spinal cord. An invertebrate does not have a spinal column.**

4. What is the difference between radial symmetry and bilateral symmetry? **Radial symmetry is when the top half of an animal looks like the bottom half. It is a type of symmetry found only in aquatic animals. Bilateral symmetry is when the right half of an animal looks like the left half.**

5. What is a gut? **It is an internal structure that digests and processes food.**

6. Which one of these is not a feature of the organism of Porifera? Sessile, filter feeders, body full of pores, bilaterally symmetric. **Sponges display all those features except for bilateral symmetric.**

7. What is a cnidocyte and what does it contain? **It is a specialized stinging cell found only in cnidarians. It contains a special organelle called the nematocyst, which stings.**

8. True or False? Some of the organisms from Platyhelminthes are parasites and others are able to live on their own (free-living). **True.**

9. True or False? Earthworms have a cerebral ganglion (worm brain) that is connected to a nerve cord. **True.**

10. What type of digestive tract do annelids have, one way or two way? **Two way.**

11. True or False? Almost all mollusks have a closed circulatory system. **False. Almost all mollusks have an open circulatory system.**

12. What molecule is an insect's exoskeleton made from? **Chitin.**

13. Why do insects molt? **Their exoskeletons are rigid and do not allow for growth. As the insect grows, it sheds its exoskeleton so it can continue to grow.**

14. What are the three parts that divide an insect's body? **Head, thorax, and abdomen.**

15. Metamorphosis that includes a larval and pupa stage is an example of complete or incomplete metamorphosis? **Complete.**

16. What is a water vascular system? **It is a series of connected tubes throughout an echinoderm's body that allow water to move through them.**

17. True or False? All members of Chordata have an external skeleton. **False. All members of Chordata have an endoskeleton.**

18. True or False? Some organisms in Chordata have a skeleton made of bone and others have one made of cartilage. **True.**

19. What is the difference between ectothermia and endothermia? **Endothermic animals perform enough chemical reactions that they are able to generate their own body heat and maintain a constant temperature independently. Ectothermic animals are not able to do this and must rely on the temperature of their environment to maintain a stable body temperature.**

20. In which type of blood vessel does gas exchange occur? **Capillaries.**

21. True or False? Gas exchange can occur in tissues, gills, or lungs. **True.**

22. Describe the blood flow in a one-loop circulatory system. **One-loop systems are only found in aquatic animals. The blood is pumped from the heart to the gills, then continues to move from the gills to the body tissues. It continues to flow back to the heart and completes the single loop.**

23. True or False? Amphibians have a three-chambered heart. **True.**

24. What is a lateral line? **The lateral line is a group of nerve cells running down both sides of the organism; they collect information from the environment such as vibration and electrical currents. This information is carried to the brain through the dorsal spinal cord. This is a sensitive part of the nervous system, which allows the organism to sense and respond to changes in its environment.**

25. Describe the metamorphosis of a frog. **The larval form of a frog is called a tadpole. Frog tadpoles live completely in the water. During the metamorphosis, the tadpole loses its tail and gills, growing legs and lungs. In addition, the circulatory system changes from a one-loop system to a two-loop system. The tadpole heart changes from a two-chambered heart into a three-chambered frog heart.**

26. True or False? Amphibians lay hard-shelled eggs that hold up to tough conditions. **False. Amphibians lay soft eggs that dry out easily.**

27. What is the structure and the importance of the amniotic egg? **It is a hard-shelled container that protects the developing embryo. The amniotic egg contains food for the developing organism and disposes of wastes. It decreases the dependence on water that amphibians have for their eggs to survive.**

28. What are the only two groups of endothermic organisms? **Mammals (Mammalia) and birds (Aves).**

29. True or False? Air sacs help to make birds lighter and allow them to perform gas exchange continuously on inhalation and exhalation. **True.**

30. What types of eggs do birds lay? **Amniotic eggs.**

31. True or False? There are three general types of birds—perching, flightless, and water birds. **False. There are four. The fourth category is birds of prey.**

32. True or False? Some mammals, all birds, and all reptiles lay amniotic eggs. **True.**

33. Do marsupials give birth to full term young or immature young? **The marsupial young are born at an immature stage.**

34. What is a joey? **A baby marsupial.**

35. What is a placenta? **It is the organ that forms between mother and developing baby to provide nutrition and waste disposal for the baby.**

36. True or False? An elephant is a type of ungulate. **False. Ungulates are animals with hooves and elephants do not have them.**

37. True or False? The animals in Edentata have no teeth but do have other specialized structures to help them obtain their food. **True.**

38. What are the only mammals that can fly? **The bats—order Chiroptera.**

39. What are pinnipeds commonly known as? **Walruses, sea lions, and seals.**

40. True or False? Primates are the only organisms to have nails instead of claws. **True.**

LIFE SCIENCE TEST #8 (Chapters 17-18)

1. True or False? In order to control the bodily functions, the nervous system needs to be able to communicate with all tissues. **True.**

2. True or False? Neurons are designed to be able to send and receive electrical impulses. **True.**

3. What are the three basic parts of a neuron? **The cell body, axon, and dendrites.**

4. True or False? Sensory neurons carry information from the spinal cord to the muscles. **False. Sensory neurons carry information from the skin and other organs to the spinal cord and brain.**

5. What does an interneuron do? **It carries information from one nerve to another.**

6. True or False? The two basic components of the nervous system are the peripheral and the central nervous system. **True.**

7. What are the meninges? **They are the tissue that covers the brain and spinal cord to help protect them.**

8. True or False? The brain is not structured so an area of the brain controls specific actions or movements and different areas of the brain have different functions in all people. **False. The brain functions have been well-mapped and are consistent from person to person.**

9. True or False? Unlike the brain, the white matter of the spinal cord is located near the outside and the gray matter is located on the inside. **True.**

10. What type of information do peripheral nerves carry and where does it go? **The peripheral nerves carry information for the muscles to contract or relax from the spinal cord to muscles. They also carry sensory information from muscles and skin to the spinal cord.**

11. What is a reflex arc and what does it consist of? **Reflexes are "built in" connections between sensory nerves, the spinal cord, and motor nerves. A reflex arc consists of a sensory nerve, an interneuron in the spinal cord, and a motor nerve.**

12. Describe the mechanism of hearing. **Sound enters the auditory canal and hits the eardrum. The sound causes the eardrum to vibrate. When the eardrum vibrates, so do the ear bones. The attachment of the middle ear bones to the cochlea causes a membrane on the cochlea to vibrate. The fluid in the cochlea moves when the membrane is vibrated by the moving of the middle ear bones. The movement of the fluid vibrates tiny hairs lining the cochlea. These hairs are connected to the nerves that transmit sound to the brain. Whenever the hairs move, information is sent to the temporal lobes of the brain, and the sound is interpreted.**

13. Does a beam of light pass through the cornea or the retina first as it passes from the front of the eye to the back of the eye? **It passes through the cornea first.**

14. True or False? Certain areas of the tongue are more sensitive to tastes than others. **True.**

15. True or False? We can smell things because the chemical vapors diffuse into the brain and stimulate the brain tissue. **False. We have receptors in our sinuses that receive certain smell molecules. When a molecule fits into a receptor, we smell that smell.**

16. What is a hormone? **Hormones are chemicals made in one tissue that are released into the blood stream and circulate to other tissues where they cause an effect on that organ's function.**

17. Describe the way that the endocrine system is controlled. **If an organ is not as active as it should be, the hypothalamus increases its activity by telling the pituitary gland**

to release a stimulating hormone. The stimulating hormone causes the activity of the target organ to increase. If an organ is over-active, then the hypothalamus tells the pituitary to stop releasing the stimulating hormone, so the target organ can shut down for a while.

18. True or False? People with type I diabetes produce too much insulin. **False. They do not produce any insulin.**

19. True or False? Ligaments connect muscle to bones. **False. Tendons connect muscles to bones.**

20. What is keratin? **It is the protein that makes up skin cells and gives skin its toughness.**

21. What is the difference between skeletal and smooth muscle? **Skeletal muscle is under active control, smooth muscle is not. Skeletal muscle has a banded appearance when viewed under the microscope, smooth muscle does not.**

22. What is a bone cell called? **An osteocyte.**

23. What is the difference between a fixed and moveable joint? **Fixed joints do not move and moveable joints do.**

24. What are the substances we need to survive that our digestive system absorbs? **Nutrients.**

25. What is another name for the digestive system? **The GI tract.**

26. True or False? Saliva provides mechanical digestion. **False. Saliva provides chemical digestion.**

27. Where does the food go after the stomach has broken it into chyme? **The small intestine.**

28. Which structures are larger, villi or microvilli? **Villi.**

29. What is blood with low oxygen concentration called? **Deoxygenated blood.**

30. What are the names of the heart valves? **Tricuspid, pulmonic, mitral, and aortic.**

31. True or False? The right ventricle generates more pressure when it contracts than the left ventricle. **False.**

32. True or False? Blood is composed of red blood cells, white blood cells, and plasma. **True.**

33. What occurs in an alveolus? **Gas exchange.**

34. What is the phrenic nerve? **It is the nerve that carries the impulse to breathe from the brain to the diaphragm.**

35. All the following are parts of the excretory system except: nephron, intron, kidney, or ureters. **Intron.**

36. What is a pathogen? **It is an organism that can cause a disease (or infection).**

37. What happens when the barrier function of the immune system fails? **The active function of the immune system kicks in and we start to actively fight the pathogen.**

38. What do antibodies do? **They stick to and coat pathogens, targeting them for destruction by cells of the immune system.**

39. True or False? Passive immunity develops after we fight off an infection. **False. Active immunity develops after an infection or an immunization.**

40. Name two conditions that prompt doctors to purposely prescribe medications to suppress a person's immune system so it cannot work well. **After organ transplantation and autoimmune disease.**

LIFE SCIENCE TEST #9 (Chapter 19)

1. What is an ecologist? **A scientist who studies ecology.**

2. How "thick" is the biosphere? **The biosphere encompasses an area from about six miles above sea level to seven miles below sea level, so the total area is thirteen miles.**

3. How are ecosystems defined? **The boundaries are defined by the researcher who is studying it.**

4. What is the difference between biotic and abiotic mass in an ecosystem? **Biotic mass is defined as all living organisms in an ecosystem. Abiotic mass is everything in the ecosystem that is not alive.**

5. True or False? Abiotic components of an ecosystem can include the water cycle, a rock, soil conditions, temperature, and the amount of sunlight an ecosystem receives. **True.**

6. What are the similarities and differences between percolation, precipitation, evaporation, and transpiration? **These are all terms that relate to movements of water in the water cycle. Percolation is the movement of water down into the ground. Precipitation is water that falls from the sky to the ground (the water can be in any form). Evaporation is water that turns into a gas (water vapor) from a body of water. The water vapor then returns to the atmosphere. Transpiration is the process of plants losing water into the atmosphere when they are photosynthesizing. This is kind of like evaporation from plants.**

7. True or False? The biotic components of an ecosystem are usually grouped into communities, populations, and individuals. **True.**

8. Which is the most accurate way to depict the energy transfer relationships that exist in ecosystems? **A food web.**

9. What is the condition called when a prey species is shaped like a leaf? **Camouflage.**

10. What is the term that describes a harmless species resembling a harmful one? **Mimicry.**

11. What is a large area of the earth that has similar temperatures, soil conditions, and plant and animal life called? **A biome.**

12. What is the process an unpopulated area of land undergoes to become an ecosystem? **The first species that move in are called pioneer species. These are plants that are usually weeds, lichens, and other species able to live in nutrient-poor soil. Pioneer species begin to loosen up the soil and provide increasing organic matter. The first plants and animals into the area form primary succession. When more fertile soil is present, larger species of plants and animals move in during the process of secondary succession. When the flora and fauna of an ecosystem are stable, then the climax community has been reached.**

PARENT COMPANION

1 | Introduction

1.0 CHAPTER PREVIEW

- Define the focus of life science as a study.
- Discuss life science as a scientific pursuit. which follows the scientific method.
- Define and discuss the scientific method.
- Define International Units (SI) and understand where they fit into scientific experimentation.
- Discuss the common methods of acquiring data during scientific experiments.

1.1 OVERVIEW

- Life science is the study of biology. Biology is the study of living organisms.
- Biologists study biology.
- Studying life science allows us to understand the world in which we live.

Topic question:

What do biologists study? **Living organisms.**

1.2 LIFE SCIENCE AS "SCIENCE"

- A scientific study follows well-established guidelines, called the scientific method, for testing hypothesis statements.
- A key component to testing a hypothesis is setting up good experiments that will test the hypothesis. Controlled experiments are excellent ways of doing this.

Topic questions:

What is the process of the scientific method? **A scientist or student develops a question regarding observations made about anything. The question is formulated into a testable hypothesis, then experimentation begins to test the hypothesis. As data is collected during experimentation, it is analyzed to test whether the data proves the hypothesis True or False. If the data supports the hypothesis, further experiments are designed and carried out to test the hypothesis more. If the hypothesis is not supported by the data, the hypothesis must either be re-formulated or thrown out altogether, and a new hypothesis statement made. If a hypothesis holds up over many years and experiments, it is considered a theory.**

True or False? The experimental group is exposed to the variable in a controlled experiment. **True.**

1.3 SCIENTIFIC MEASUREMENTS

- All scientific measurements are obtained by using SI units, or the International System of units.

Topic question:

What is the difference between a measured and a derived unit? **Measured units can be measured directly using some measuring device; derived units must be calculated because they represent mathematical relationships between two or more measured or other derived units.**

1.4 METHODS OF OBTAINING DATA

- The microscope is one of the most common instruments used to obtain data during scientific experiments.

- Electron microscopes are more powerful than light microscopes and can magnify images many thousand times more than a light microscope.

- CT and MRI are becoming more frequently utilized by biologists because they allow for examination of detail without destroying tissue or samples.

Topic question:

List at least five ways data can be obtained during an experiment. **Light microscopy, electron microscopy, thermometer, scales, speedometers, cameras, unaided vision, hearing, taste, touch, smell, CT scanning, MRI, and x ray (to name a few).**

1.5 KEY CHAPTER POINTS

- Life science is the study of living organisms.

- The balance of all living organisms on the planet is fragile, yet strong, at the same time. Studying the life sciences allow us to appreciate this fact better.

- Life science research is subject to the strict principles of the scientific method.

- The scientific method involves making a hypothesis statement, making observations through experimentations, and analyzing the results to test whether a hypothesis is possibly true or not.

- A hypothesis statement that has stood the test of time is called a theory.

- SI units are standard units of measure that all researchers use when collecting experimental data.

- Light microscopes, electron microscopes, MRI and CT scanners, and field studies are all methods that can be used to test a hypothesis.

2 | Characteristics of Life

2.0 CHAPTER PREVIEW

In this chapter we will:

- Investigate the meaning of "organism."
- Define the properties that all living things share.
- Discuss the common things that almost all organisms on earth require to live.

2.1 OVERVIEW

- All living things share certain properties associated with being alive. This chapter will discuss them.

Topic question:

True or False? In order to study life, one must know what is alive and what is not. **True. Establishing common criteria to define life allows for accurate study and discussion of the experimentation process.**

2.2 PROPERTIES OF LIFE: CELLS

- The basic functional unit of all life forms is the cell. All living things are made up of one or more cells.

Topic question:

True or False? A multicellular organism is made up of two or more cells. **True.**

True or False? Unicellular organisms are not alive because they only contain one cell. **False. Unicellular organisms are every bit as alive as multicellular organisms.**

2.3 PROPERTIES OF LIFE: DEOXYRIBONUCLEIC ACID (DNA)

- All living things contain DNA.
- The basic unit of DNA is the gene.
- DNA (genes) are passed from parent to offspring.

Topic question:

Why is DNA called the "blueprint of life?" **DNA contains the genetic information that codes for, or controls, the development of every aspect of an organism's appearance and function.**

2.4 PROPERTIES OF LIFE: REPRODUCTION

- All organisms reproduce.
- There are two forms of reproduction—sexual and asexual.
- Asexual reproduction involves cell division without mixing of chromosomes. Sexual reproduction involves reproduction through the combination of chromosomes from a male and female organism to form a new organism.

Topic question:

What is the difference between sexual and asexual reproduction? **Sexual reproduction occurs by the DNA of two organisms combining to form a new organism. Asexual reproduction occurs without combining genetic information. Sexual reproduction forms organisms that are genetically different than either parent, but asexual reproduction leads to organisms that are genetically identical to the parent cell.**

2.5 PROPERTIES OF LIFE: COMPLEX AND ORGANIZED

- All life forms are highly organized and complex structures.

Topic questions:

True or False? Even unicellular organisms are complex and organized. **True. All life forms are complex and organized no matter how big or small they are.**

Do multicellular organisms have a higher degree of complexity than single cell organisms? **Of course. They are larger and have more cells, so they are naturally more complex and organized relative to unicellular organisms.**

2.6 PROPERTIES OF LIFE: RESPONSIVE

- All living things are responsive to their surroundings or environment.
- The ability to respond is controlled by different types of receptors.

Topic question:

What are the structures called that allow an organism to sense and respond to their environment? **Receptors.**

2.7 PROPERTIES OF LIFE: ENERGY EXTRACTION AND USAGE

- All organisms extract energy they need to survive from their environment.
- Producers extract the energy they need from the sun through photosynthesis.
- Consumers and decomposers extract the energy they need by eating other organisms and using it during cellular respiration.

Topic questions:

True or False? Plants convert the sun's energy through the process of cellular respiration. **False. Plants convert the sun's energy through photosynthesis.**

True or False? Plants extract the energy they need from their environment, but animals do not. **False. All living things extract the energy they need to live from their environment.**

2.8 PROPERTIES OF LIFE: HOMEOSTASIS

- All organisms maintain a stable internal environment, a process called **homeostasis**.

Topic question:

When a fish that is cold swims into warmer water to warm up, then later swims to cooler water to cool down, what property of life is the fish displaying? **Homeostasis.**

2.9 PROPERTIES OF LIFE: GROWTH

- All living things grow.

Topic questions: none

2.10 PROPERTIES OF LIFE: CLASSIFICATION

- All living things are systematically classified based on similarities and differences to other organisms.
- The scientific pursuit of classifying organisms is called taxonomy.
- The current taxonomical system is a seven-level system founded on that of Carl Linnaeus.

Topic question:

What is a binomial name? **It is a specific name for an organism that no other organism shares. The binomial name is a unique identifier. The first part of a binomial name is the genus the organism is classified into; the second name is the species.**

2.11 PROPERTIES OF LIFE: COMMON THINGS ALMOST ALL ORGANISMS NEED

- All living organisms need food for energy and a habitat in which to live. In addition, most organisms require oxygen, but not all. There are some organisms that die in the presence of oxygen

Topic questions:

What is an aerobic organism? **One that needs oxygen to survive.**

True or False? Air is made up mainly of nitrogen, oxygen, carbon dioxide, and argon. **True.**

2.12 KEY CHAPTER POINTS

- All living things have the following properties in common:
    - Composed of one (unicellular) or more cells (multicellular).
    - Contain the blueprint of life, DNA.
    - Sexually or asexually reproduce to make more cells and more organisms.
    - Complex and organized on many levels.
    - Have ways to sense and respond to changes in their environment.
    - Extract energy from their surroundings.
    - Maintain homeostasis.
    - Grow in size.
    - Can be classified based on similarities and differences in structure and function.
    - Almost all organisms need water, air, food, and a habitat to live.

The Chemistry of Life
3 and Light Energy

3.0 CHAPTER PREVIEW

- Investigate the concepts of matter, mass, and weight.
- Define the structure of atoms, elements, and molecules.
- Discuss the structure of the organic molecules: carbohydrates, proteins, lipids, and nucleic acids.
- Discuss the properties of the electromagnetic spectrum and visible light.

3.1 OVERVIEW

- Atoms link together to form molecules; molecules link together to form the complex structures that are cells and organisms.

Topic question: none

3.2 MATTER

- Matter is anything that takes up space and has mass.
- Mass is the amount of matter something has.
- Mass and weight are not the same.

Topic questions:

Why are matter, mass, and weight not the same thing? **Matter is anything that takes up space. It is the physical presence of atoms and molecules. Mass is the amount of matter that something has. Unless part of the object is broken off of the original, mass does not change (under "normal" conditions). Weight is dependent on the pull of gravity that is acting on the object that contains matter and has mass. Weight changes depending on the gravitational pull, but matter and mass do not.**

What is one way that you can prove something you cannot see, such as a gas, contains matter? **Blow up a balloon. If the gas you were blowing into the balloon did not have matter, then the balloon would not fill. The matter in the gas takes up space inside the balloon, and the more matter you blow into the balloon, the fuller it gets.**

3.3 ATOMS AND MOLECULES

- The smallest building block of matter is an atom. Every atom is made up of protons, neutrons, and electrons.
- An element is a special type of molecule. An element is matter composed of only one type of atom.
- Two or more atoms link together to form molecules.

Topic question:

True or False? A molecule is matter composed of only one type of atom. **False. A molecule is made up of two or more atoms linked together. An element is matter composed of only one type of atom.**

3.4 ORGANIC MOLECULES

- Organic molecules are molecules that contain carbon and are normally made by living cells.
- There are four classes of organic molecules: carbohydrates, lipids, proteins, and nucleic acids.

Topic questions:

What is an organic molecule? **Molecules that contain carbon and are normally made by living cells.**

What are the four types of organic molecules? **Carbohydrates, lipids, proteins, and nucleic acids.**

3.5 CARBOHYDRATES

- Carbohydrates are molecules made up of carbon, oxygen, and hydrogen atoms; they are commonly called "sugars."
- There are two types of carbohydrates—simple and complex.

Topic question:

What three atoms compose all carbohydrate molecules? **Carbon, hydrogen, and oxygen.**

True or False? Glucose is an example of a complex carbohydrate. **False. Glucose is a simple carbohydrate.**

3.6 LIPIDS (FATS)

- Lipids are also called fats or oils.

Topic question:

True or False? Lipids are usually solid at room temperature, while fats are usually liquid. **False. Fats are usually solid and oils are usually liquid.**

3.7 PROTEINS

- All proteins are built by linking together smaller units called amino acids.
- The cell's structure is formed almost entirely by lipids and proteins.
- Special proteins called enzymes control the chemical reactions of the cell.

Topic questions:

How are proteins made? **By linking together smaller units called amino acids.**

True or False? Proteins are not sensitive to changes in their amino acids, so the amino acids can be changed in a protein without much effect, if any, to the organism or the way the protein works. **False. Proteins are sensitive to small changes in their amino acids, so much so that sometimes just changing one amino acid can make a protein or enzyme completely non-functional.**

3.8 NUCLEIC ACIDS

- Nucleic acids are made up of smaller units called nucleotides.
- There are three important nucleic acids we will learn about: DNA, RNA, and ATP.

3.9 LIGHT ENERGY

- Sunlight is the ultimate source for all the energy production on earth. Without the sun, producers would have no way to make glucose, and consumers would then have no glucose to eat.
- Radiant energy, also called the electromagnetic spectrum, is the entire range of radiant energies or wave frequencies from the longest to the shortest wavelengths emitted by the sun.
- The visible light portion of the sun's radiant energy is used by plants to make glucose during photosynthesis.
- The visible spectrum consists of all the separate colors seen when white light is passed through a prism.
- Objects appear the color they are because the light they reflect contains radiant energy in that particular wavelength.
- Plants can only use the sun's energy they absorb.

Topic questions:

Why is it said that the sun is the ultimate source of all biological energy on earth? **Without the sun, plants would not perform photosynthesis, so they would not make any glucose molecules. Without glucose molecules, neither producers nor consumers would have anything to use during cellular respiration. Without cellular respiration, no organisms would be able to make ATP, and without ATP, no chemical reactions would happen. Life would end on earth without the sun.**

True or False? Plants use the energy they reflect for photosynthesis. **False. Plants can only use the energy they absorb. Reflected light energy is lost light energy.**

True or False? Objects appear the color they are because of the light they reflect. **True.**

3.10 KEY CHAPTER POINTS

- Matter is anything that has mass. Mass is the amount of matter that an object contains. The weight of an object depends on the gravitational pull acting on the object.

- The basic unit of matter is the atom. Two or more atoms bonded together form a molecule.

- The molecules of life are carbohydrates, lipids, proteins, and nucleic acids.

- The sun emits two types of energy—heat and the electromagnetic spectrum.

- Making the molecules of life depends on the energy contained in the waves of the visible light part of the electromagnetic spectrum.

- All objects reflect and absorb different wavelengths of visible light. Objects appear a certain color because of the wavelength of light(s) they reflect.

- Plants use the energy from the wavelengths of visible light they absorb to make glucose in a process called photosynthesis.

Cell Membrane:
4 | Passive and Active Transport

4.0 CHAPTER PREVIEW

In this chapter we will:

- Investigate the history and meaning of cell theory.

- Define the properties of cells.

- Investigate the differences between the two basic cell types that make up all organisms— eukaryotic cells and prokaryotic cells.

- Describe the structure and function of the cell membrane.

- Understand the passive and active processes used to move substances across a membrane.

4.1 OVERVIEW

- The outer boundary of the cell is formed by the cell membrane.

- All biological membranes have a similar structure.

Topic question:

What is the outer boundary of the cell called? **The cell membrane.**

4.2 CELL THEORY

- Cell theory states that all organisms are composed of one or more cells. It also states the cell is the basic, functional unit of life.

- The cell theory further states that only cells can make other cells.

- Organisms that are made up only of one cell are called unicellular. Organisms that are more than one cell are called multicellular.

Topic questions:

True or False? Cell theory states there is no smaller unit capable of life than a cell. **True.**

True or False? Cells are made up of non-living molecules. **True.**

4.3 PROPERTIES OF CELLS

- Cells are diverse with many different shapes, sizes, and functions.

Topic question:

Why are cells described as spheres or cubes rather than as circles and squares? **Because cells are three-dimensional structures and should be described with three-dimensional words such as *cube* and *sphere*, rather than two-dimensional words such as *circle* and *square*.**

4.4 FUNCTION OF A CELL

- Since the cell is the basic functional unit of life, the function of the cell is the same as the properties of life.

- One of the properties of cells discussed in detail is the organization of cells into tissues, tissues into organs, and organs into organism systems that are seen in multicellular organisms.

Topic questions:

What do you call a group of cells working together with a similar function? **A tissue.**

What is an organ system? **Two or more organs working together to accomplish a similar function for the good of the organism.**

4.5 BASIC CELLULAR STRUCTURE

- All cells have the basic cell structure of a cell membrane enclosing cytoplasm, organelles, and DNA.
- Prokaryotic cells do not have their DNA enclosed in a nucleus and do not have membrane-bound organelles.
- Eukaryotic cells have membrane-bound organelles, including a nucleus, which contains the DNA.
- Many organisms have an additional layer outside the cell membrane called a cell wall, but no animal cells have a cell wall.

Topic questions:

What is a prokaryote? **An organism that is composed of a prokaryotic cell.**

How does a prokaryotic cell differ from a eukaryotic cell? **A prokaryote does not have a nucleus, nor does it have membrane-bound organelles. Eukaryotes have both.**

What is a nucleoid? **The area of the cytoplasm where prokaryotic DNA is contained.**

4.6 CELL MEMBRANE

- All cell membranes are composed of special lipid molecules called phospholipids.
- Phospholipids organize into a lipid bilayer in watery environments.
- The cell membrane also contains proteins.
- The cell membrane is selectively permeable.

Topic questions:

Why do phospholipids orient with their tails facing toward and their heads facing away from one another? **All cells are surrounded by a watery environment. The phospholipid heads like water, and the tails do not. When many phospholipid molecules are placed into a watery environment, the tails naturally align toward the center of the lipid bilayer to get as far from the water as they can. The heads do not mind at all sitting out in the water, so they face outward, toward the water.**

What is meant by "the membrane is selectively permeable?" **It means only certain molecules can get across—the membrane is picky and does not let just any molecule cross into or out of the cell. There are some molecules able to freely pass through the membrane, but most cannot.**

4.7 CELL MEMBRANE: DIFFUSION

- Diffusion is the movement of molecules from an area where they are in high concentration to an area where they are in low concentration. Molecules pass through the cell membrane by diffusion when the membrane is permeable to them.

Topic questions:

True or False? Diffusion is the random movement of molecules from an area of high concentration to an area of low concentration. **True.**

What is meant when it is said that equilibrium has been established relative to diffusion. **It means the concentration of molecules is equal on each side of the membrane.**

4.8 CELL MEMBRANE—OSMOSIS

- Osmosis is a specific type of diffusion—it is the diffusion of water across a selectively permeable membrane.

Topic question:

What is osmosis? **It is the diffusion of water across a selectively permeable membrane from an area of high water concentration to low concentration.**

4.9 CELL MEMBRANE: PASSIVE AND ACTIVE TRANSPORT

- Passive transport is a mode of moving molecules across a membrane that does not require the cell to use energy.
- Diffusion, osmosis, and gated channel transport are all forms of passive transport.
- Active transport requires the cell to use energy to move molecules across a cell membrane.
- Cell pump proteins, exocytosis, and endocytosis are all forms of active transport.

Topic questions:

What is the difference between active and passive transport? **Passive transport occurs without the cell using any energy. Active transport requires the cell to use energy.**

True or False? Cell pump proteins actively move molecules from areas of high concentration to areas of low concentration. **True.**

4.10 CELL MEMBRANE: ENDOCYTOSIS AND EXOCYTOSIS

- Endocytosis and exocytosis are both modes of active transport that move molecules by enclosing them in a membrane.

Topic question:

True or False? Endocytosis expels items from the cell and exocytosis brings them into the cell. **False. Endocytosis brings molecules into the cell and exocytosis expels them from the cell.**

4.11 KEY CHAPTER POINTS

- The cell theory states that the cell is the basic functional unit of life.
- Unicellular organisms are made up of one cell, and multicellular organisms are made up of two or more cells.
- Cells are different in their structure and function.
- The function of a cell is exactly the same as the properties of life.
- There are two types of cells—prokaryotic and eukaryotic. The basic cell structure for both cell types is a cell membrane that encloses cytoplasm and organelles.
- Eukaryotic cells have more types of organelles than prokaryotes. Eukaryotic organelles are membrane bound, prokaryotic organelles are not. Eukaryotic DNA is contained in the nucleus, prokaryotic DNA is in the cytoplasm.
- The cell membrane is made mainly of phospholipids and proteins.
- The cell membrane moves substances into and out of the cell with diffusion, osmosis, passive transport, active transport, endocytosis, and exocytosis. The first three do not require any energy, the last three do.

5 | The Cell Interior and Function

5.0 CHAPTER PREVIEW

In this chapter we will:

- Discuss the structure and function of the common eukaryotic organelles:
  - cytoplasm
  - nucleus
  - nucleoplasm
  - nucleolus
  - ribosome
  - endoplasmic reticulum
  - Golgi apparatus
  - lysosomes
  - mitochondria
  - vacuoles
  - chloroplasts

- Investigate the set of chemical reactions plants perform to make glucose. This process is called photosynthesis.

- Discuss the set of chemical reactions plant and animals use to make their energy molecules. This process is called cellular respiration.

- Discuss the relationship between photosynthesis and cellular respiration.

5.1 OVERVIEW

- This chapter, we will be discussing the individual units inside of the cell, called organelles.

Topic question:

What is an organelle? **It is the individual functional unit of the cell. Every type of organelle has a specific function in the cell.**

5.2 REVIEW OF PROKARYOTIC AND EUKARYOTIC CELLS

- A review or prokaryotic cells and eukaryotic cells is discussed.

Topic question:

If you were looking through a microscope at a cell with a nucleus, what type of cell would it be: prokaryotic or eukaryotic? **Eukaryotic.**

5.3 ORGANELLES: WHY DO WE HAVE THEM?

- Organelles exist for enhanced control over the function of the cell and to distribute the burden of function because eukaryotic cells perform too many functions for one organelle to be able to handle it.

Topic question:

What are two differences between a plant and animal cell? **Plant cells have cell walls and chloroplasts, animal cells do not.**

5.4 CYTOPLASM

- Cytoplasm is the jelly-like fluid inside a cell. It contain water, organelles, proteins, ions, and other molecules.

Topic question:

Why is it important that the cytoplasm is somewhat watery? **The fluid nature of the cytoplasm allows for wastes, molecules, and organelles to move around in the cytoplasm.**

5.5 EUKARYOTIC ORGANELLES: NUCLEUS

- The nucleus contains the DNA, the nucleolus, and nucleoplasm.

Topic questions:

What is the nucleolus and what does it do? **It is an area in the nucleus that manufactures pieces of ribosomes.**

What are nuclear pores? **They are small openings in the nuclear membrane that allow molecules to pass into and out of the nucleus.**

5.6 EUKARYOTIC ORGANELLES: RIBOSOMES

- Ribosomes are organelles that make proteins.
- Ribosomes in eukaryotes and prokaryotes do not have membranes.

Topic questions:

Do ribosomes have a membrane? **No, they are the only non-membrane-bound organelle in a eukaryotic cell.**

What do ribosomes do? **They make proteins using instructions contained in mRNA.**

5.7 EUKARYOTIC ORGANELLES: ENDOPLASMIC RETICULUM

- The endoplasmic reticulum, or ER, is an extensive network of folded membranes and sacs in the cytoplasm.
- Rough ER looks bumpy through the electron because there are ribosomes attached to it.
- Smooth ER looks smooth when looked at through an electron microscope because there is nothing attached to it.

Topic question:

True or False? ER packages and transports proteins after they are made by ribosomes. **True.**

5.8 EUKARYOTIC ORGANELLES: GOLGI APPARATUS

- The Golgi apparatus is made up of a series of flat, stacked tubes with little sacs at the end of the tubes.
- The Golgi apparatus receives many proteins from the ER so the Golgi can store or process and transport the protein.

Topic question:

What is the name for the series of flattened and stacked tubes with small sacs at the end of them? **The Golgi apparatus.**

5.9 EUKARYOTIC ORGANELLES: LYSOSOMES

- Lysosomes are membrane-bound garbage disposers of the cell.

Topic question:

Where are lysosomes made? **In the Golgi apparatus.**

5.10 EUKARYOTIC ORGANELLES: MITOCHONDRIA

- Mitochondria are the organelles where the chemical reactions occur that make the cell's fuel. For that reason, mitochondria are often called the "powerhouses" of the cell.

Topic questions:

What are the inner folds of membranes called in mitochondria? **Cristae.**

What happens in the cristae? **They are where ATP is made.**

5.11 EUKARYOTIC ORGANELLES: VACUOLES

- Vacuoles are storage organelles.

Topic question:

What is the function of a vacuole? **It is for storage of substances in the cell.**

5.12 EUKARYOTIC ORGANELLES: CHLOROPLASTS

- Chloroplasts are found in algae and plants; animal cells do not contain them.
- Chloroplasts perform photosynthesis.

Topic question:

What important photosynthetic molecule do chloroplasts contain? **Chlorophyll.**

5.13 PLANT AND ANIMAL ENERGY

- All processes that cells perform require energy.
- The energy that cells need to perform their chemical reactions is obtained from ATP.
- Cellular respiration, occurring in mitochondria, is the biochemical set of reactions that makes ATP using energy released when glucose is broken down.
- Photosynthesis, occurring in chloroplasts, is the biochemical set of reactions that capture the sun's energy and use it to make glucose from carbon dioxide and water.

Topic questions:

True or False? ATP is the molecule made during photosynthesis. **False. ATP production is a function of mitochondria.**

What is important about the glucose plants make during photosynthesis? **The glucose provides all the available glucose for cellular respiration to occur in animals and other consumers.**

5.14 PHOTOSYNTHESIS: FUNCTION OF CHLOROPLASTS

- Photosynthesis is the set of chemical reactions occurring in chloroplasts in which the sun's energy is captured and used to make glucose from carbon dioxide and water.
- Plants and algae perform photosynthesis.

Topic question:

True or False? During photosynthesis, six molecules of water and six molecules of carbon dioxide are used to make one molecule of glucose and six molecules of oxygen. **True.**

5.15 CELLULAR RESPIRATION

- Cellular respiration is the series of chemical reactions occurring in mitochondria in which glucose is broken down, then the released heat is captured to make molecules of ATP.
- Almost all organisms on earth perform cellular respiration, and they all need glucose, manufactured in plants, to perform it.
- ATP is the universal power molecule that cells use to fuel their chemical reactions.

Topic questions:

What generally occurs during cellular respiration? **Glucose is broken down by mitochondria. This releases energy, which is captured and used to make many molecules of ATP.**

What is important about ATP? **It is the molecule that almost every cell on earth uses for energy.**

What is the relationship between cellular respiration and photosynthesis? **Cellular respiration releases carbon dioxide into the air, which plants need to use during photosynthesis. Photosynthesis releases oxygen into the air, which is needed during cellular respiration. Also, photosynthesis produces glucose, which is broken down during cellular respiration to make ATP.**

5.16 ANAEROBIC CELLULAR RESPIRATION

- Anaerobic cellular respiration is the process of making ATP when oxygen is not present. It is a faster way to make ATP, but much less efficient.

Topic questions:

True or False? When oxygen supply is low, cells begin to make ATP through fermentation. **True.**

True or False? Fermentation makes more ATP per molecule of glucose than aerobic cellular respiration. **False. Fermentation makes it quicker, but much less of it per molecule of glucose than aerobic respiration.**

5.17 KEY CHAPTER POINTS

- Organelles are the individual functional units of the cell.
- All eukaryotic organelles, except ribosomes, are surrounded by membranes.
- The nucleus houses and protects the DNA.
- Ribosomes manufacture proteins.
- Endoplasmic reticulum processes and packages proteins and transports molecules through the cell.
- Golgi apparatus further processes proteins and also makes lysosomes.
- Lysosomes contain enzymes that break molecules and substances down.
- Mitochondria manufacture ATP through cellular respiration.
- Vacuoles store substances.
- Chloroplasts make glucose through photosynthesis.
- Photosynthesis is the biological process of manufacturing glucose from water and carbon dioxide using energy provided by the sun.
- Aerobic cellular respiration is the biological process of manufacturing ATP using the energy released during the metabolism of glucose in the presence of oxygen.
- Anaerobic cellular respiration is the biological process of manufacturing ATP without oxygen.

6 | DNA Structure and Function

6.0 CHAPTER PREVIEW

In this chapter we will:

- Investigate how DNA is organized into the smaller units of chromosomes and genes.
- Define the properties of chromosomes and genes.
- Discuss the genetic code.
- Learn the pathway that is taken for a protein to be made from a gene.
- Discuss the biological process of DNA replication.
- Discuss the concept of mutations.

6.1 OVERVIEW

- In this chapter, we are going to learn more about DNA and how it is organized in the nucleus; how it instructs the ribosomes to make proteins; and how the normal sequences of DNA are changed or mutated.

Topic question: none

6.2 DNA STRUCTURE

- DNA is made from four nucleotides linked together.
- Each nucleotide has a central five-carbon sugar with a phosphate group attached to one end and a nitrogen-containing base to the other end.
- DNA forms a double-stranded helix molecule shape.

Topic question:

What are the four nucleotides of DNA? **Adenine, guanine, thymine, and cytosine.**

6.3 CHROMOSOMES

- Chromosomes are smaller, more manageable segments of DNA found in eukaryotic cells.
- Eukaryotic chromosomes are paired. Each chromosome pair contains information for the same traits (contains the same genes).

Topic questions:

What are cells that contain two pairs of each chromosome called? **Diploid cells.**

Why are chromosomes paired? **Because each chromosome of a pair contains genes for the same trait.**

6.4 AUTOSOMES AND SEX CHROMOSOMES

- Autosomes carry information that codes for the development of body characteristics, or non-sexual characteristics.
- Sex chromosomes contain information coding for the development of sexual characteristics.

Topic questions:

How many chromosomes, total, do humans have? **Forty-six.**

How many pairs of chromosomes do humans have? **Twenty-three.** If a person has an X and a Y chromosome, is that person a male or female? **Males are XY and females are XX.**

6.5 GENES

- Genes are even smaller segments of DNA contained on chromosomes.

- Each gene contains information that codes for the production of one protein. As such, each gene codes for the development of one trait.

- The genetic code is the coded information contained in the sequence of a gene's nucleotides instructing the cell how to make a protein.

Topic questions:

True or False? Chromosomes are the smallest unit of the genetic code. **False. Genes are the smallest unit of the code and are contained in chromosomes.**

True or False? Genes are segments of a chromosome that contained on opposite strands of DNA so one gene is contained on both strands of DNA. **False. A gene is on one DNA strand only. One gene is not partly contained on one strand with the rest on another strand.**

6.6 THE GENETIC CODE

- A codon is a three-nucleotide sequence of DNA (a gene) that codes for the insertion of an amino acid into a protein.

- Each amino acid has at least one codon in the genetic code.

- A gene is a string of codons that codes for the production of a protein.

Topic questions:

How many nucleotides are contained in a codon? **Three.**

True or False? A gene is a segment of DNA that contains codons. **True.**

6.7 CONCEPTUAL OVERVIEW OF MAKING A PROTEIN

- The genetic code of the gene is made into a coded molecule of mRNA in the nucleus.

- mRNA leaves the nucleus and enters the cytoplasm to carry the message to the ribosomes.

- Ribosomes read and decode the message in the mRNA. As they do, they make the protein that the codons in the mRNA message instructs them to make.

Topic questions:

Where is mRNA made? **The nucleus.**

Where are proteins made? **The cytoplasm.**

What cell organelle makes proteins? **Ribosomes.**

How do ribosomes know how to make the right protein? **They decode the message in the mRNA that tells them which amino acids to link together. Ribosomes simply follow the instructions they read in the codons of the mRNA.**

6.8 mRNA

- mRNA is made from nucleotides—adenine, uracil, cytosine, and guanine.

- The process of making mRNA from DNA is called transcription.

Topic questions:

What nucleotide base pairs with adenine in mRNA? **Uracil.**

How is a molecule of mRNA made? **Special proteins, called enzymes, unzip the DNA that needs to be made into a molecule of mRNA. This is the gene that codes for the protein that is to be made. Other enzymes make mRNA from DNA—exactly transferring the codon sequence from DNA to mRNA to preserve the genetic code.**

6.9 FROM DNA TO PROTEIN

- The process of ribosomes making proteins from mRNA is called protein synthesis or translation.

- The ribosome knows that it is inserting the correct amino acid into the protein because it links the mRNA codon to an area on tRNA. If the tRNA does not link to the mRNA codon, then the ribosomes know that tRNA molecule is not carrying the proper amino acid the code calls for.

Topic questions:

True or False? Ribosomes "read" the mRNA molecule as we would read a book—one way, codon by codon (letter by letter in the case of reading a book). **True.**

What is another word for protein synthesis? **Translation.** Where in the cell does translation occur? **The cytoplasm.**

True or False? tRNA is a molecule that shuttles amino acids to the ribosomes. **True.**

6.10 DNA REPLICATION

- DNA replication is the biological process of making an exact copy of DNA prior to cell division.

Topic questions:

True or False? When DNA is replicated, one strand is copied first, then the opposite strand is copied. **False. Both strands are copied at the same time.**

Why would a cell need to copy its DNA? **So when one cell divides into two cells, each of the two cells receives a copy of the instructions telling them how to function properly.**

6.11 MUTATION

- Mutations are errors that occur during DA replication resulting in incorrect amino acids being inserted into the DNA.

Topic questions:

Mutations are not necessarily bad for an organism. Why? **The mutation may not ever be noticed if it occurs in an area of the DNA that does not contain a gene.** True or False: there are a number of diseases that are caused by mutations. **True.**

6.12 WAYS IN WHICH DNA MUTATES

- There are two generic types of mutation—point and chromosomal.

Topic question:

Which type of mutation, point or chromosomal, involves more DNA? **Chromosomal.**

6.13 KEY CHAPTER POINTS

- DNA is composed of four nucleotides—adenine, thymine, guanine, and cytosine.

- Eukaryotic DNA is divided into smaller units called chromosomes. Chromosomes are divided into still smaller units called genes.

- One gene contains the coded information for the production of one protein.

- Transcription is the biological process of making a molecule of mRNA from DNA. mRNA takes the coded message from DNA into the cytoplasm to the ribosomes. Ribosomes decode the message and make the protein indicated by the coded message in mRNA.

- Protein synthesis is the biological process of making a protein.

- DNA replication is the biological process in which all chromosomes inside of a cell are copied before the cell divides.

- Mutations are errors that occur during DNA replication and result in the wrong sequence of nucleotides being copied from the old DNA strand to the new DNA strand.

7|Cell Reproduction

7.0 CHAPTER PREVIEW

In this chapter we will:

- Investigate the biological process of both sexual and asexual cell division.
- Define the asexual cell division processes of binary fission and budding.
- Examine mitosis in detail.
- Examine meiosis in detail.

7.1 OVERVIEW OF CELL DIVISION

- Cell reproduction (or cell division) is the process of one cell dividing into two cells.
- Cell reproduction is the process that generates more cells and more organisms.

Topic question:

What is the general process of cell division? **One parent cell divides into two daughter cells.**

7.2 OVERVIEW OF ASEXUAL CELL REPRODUCTION

- There are two forms of cell reproduction—sexual and asexual.
- Asexual reproduction does not entail combination of genetic material from two organisms, sexual reproduction does. As such, asexual cell reproduction results in the production of clones.

Topic questions:

What is the general process of cell reproduction, including what occurs to the DNA? **The DNA in the parent cell is replicated so the cell contains two exact copies of DNA. The DNA is then separated to opposite ends of the cell so one end receives a complete copy. Then the parent cell divides into daughter cells so that each of the daughter cells contains a complete copy of DNA (chromosomes).**

What is it called when the chromosomes (DNA) separate during cell division? **Karyokinesis.**

7.3 BINARY FISSION

- Binary fission is the asexual cell reproduction method that prokaryotes use to divide.

Topic question:

Describe the process of binary fission. **The prokaryote cell copies its DNA, then begins to elongate. As the cell gets longer, the two copies of DNA move farther away from one another. The cell then divides into two cells so that each daughter cell contains an exact copy of the DNA.**

7.4 BUDDING

- Budding is a common way many unicellular eukaryotes reproduce new organisms. Some multicellular eukaryotes also bud

Topic question:

True or False? When a multicellular organism buds, a mini version of the adult grows from the adult organism, then falls off. **True.**

7.5 MITOSIS

- Mitosis is the biological process of one eukaryotic parent cell dividing into two daughter cells that are genetically identical.
- The life of a cell is divided into different stages, called the cell cycle, based on what the cell is "doing."
- Mitosis is divided into stages: prophase, metaphase, anaphase, and telophase.

Topic questions:

True or False? The stages of the cell cycle are divided based on what is happening to the cell membrane during mitosis. **False. The cell cycle is divided into stages based on what is happening to the chromosomes.**

When do the chromosomes get pulled apart during mitosis? **Anaphase.**

7.6 OVERVIEW OF SEXUAL CELL REPRODUCTION

- Sexual cell reproduction is the process of forming a new organism by combining DNA (chromosomes) from a male and parent with chromosomes from a female parent.
- This requires the production of special types of cells, called gametes.
- When the sperm fuses with—or fertilizes—the egg, a new single-celled organism is formed called a zygote.
- The biological process of forming gametes is called meiosis.

Topic questions:

True of false? Haploid cells contain one copy of every chromosome in them and diploid cells contain two copies of every chromosome. **True.**

True or False? Gametes are diploid and body cells are haploid. **False. Body cells are diploid and gametes are haploid.**

Why do gametes have to be haploid cells. **So that when the chromosomes combine during fertilization, the zygote has the proper number of chromosomes. If gametes were diploid, the zygote would have twice as many chromosomes as is normal.**

7.7 MEIOSIS

- Meiosis is divided into two stages—meiosis I and II.
- The chromosomes are replicated before meiosis I only.
- Meiosis results in the formation of four haploid gametes from one diploid reproductive cell.

Topic question:

True or False? Meiosis results in the production of four haploid gamete cells from one diploid parent reproductive cell. **True.**

7.8 KEY CHAPTER POINTS

- Cell reproduction involves two forms—sexual and asexual.
- Asexual forms of cell reproduction are binary fission, budding, and mitosis.
- Asexual reproduction results in the formation of diploid cells from diploid cells that are clones of the parent cell.
- Binary fission is the form of cell division prokaryotes perform.
- Budding is the form of cell division unicellular and some multicellular eukaryotes perform.
- Mitosis is the form of cell division that multicellular eukaryotes perform.
- Meiosis is the special type of cell division that forms haploid gametes from diploid reproductive cells.

8 | Heredity

8.0 CHAPTER PREVIEW

In this chapter we will:

- Introduce and define new terms related to the study of heredity and genetics.
- Examine Gregor Mendel's experiments in heredity as an example of both excellent science and groundbreaking results.
- Discuss the concept of dominant and recessive traits and the genes that code for them.
- Learn the use of Punnett squares.
- Discuss inheritance patterns.
- Review the process of mutation as it relates to causing genetic diseases.
- Discuss recessive and sex-linked genetic diseases.
- Learn the basics of genetic engineering.

8.1 INTRODUCTION

- Every organism produced through sexual reproduction receives half of their chromosomes from their male parent and the other half from the female parent.
- Traits are controlled by genes. Since genes, contained in chromosomes, are passed from parent to child, traits are passed from parent to child.

Topic question:

What is a trait? **A characteristic of an organism.**

8.2 HEREDITY

- The passage of traits, transmitted or controlled by genes, is called heredity and is what geneticists study.

Topic question:

Geneticists study the passage of traits from generation to generation. **True**

8.3 GREGOR MENDEL

- Gregor Mendel was the first person to identify how traits are transmitted from parent to offspring. He carefully studied how various traits of the common garden pea, *Pisum sativum*, were passed from generation to generation.
- He selectively mated, or crossbred (called "crossing") pea plants with the traits he wanted to study with one another.

Topic questions:

What is the P generation? **It is the starting generation of a genetic study.**

True or False? A hybrid is an organism that produces offspring with exactly the same traits over time. **False. A hybrid is an organism that contains genes from two different purebred parents. This means they have genes that code for different traits and will produce offspring with different traits over many generations. A purebred is an organism that produces offspring with the same traits generation after generation.**

8.4 DOMINANT AND RECESSIVE TRAITS

- Mendel found there were two traits for all the characteristics he studied—one was always dominant to the other, and he called this the dominant trait. The other is termed recessive.

Topic question:

True or False? The yellow pea trait is recessive to the green pea trait. **False. The yellow pea trait is the dominant pea color.**

8.5 ALLELES AND TRAITS

- An allele is an alternate form of a gene. Alleles code for the same characteristic but slightly different traits.
- Like chromosomes, alleles are paired—one allele for a trait is contained on each chromosome.
- Purebred organisms contain two of the same allele for a characteristic.
- Hybrids contain two different alleles for a characteristic.

Topic questions:

When a dominant and recessive allele are present in the same organism, which trait is expressed and which trait is suppressed? **The dominant trait is expressed and the recessive trait is suppressed.**

If an organism has the following two alleles for a characteristic—GG—is this organism a purebred or hybrid? **This is a purebred because the alleles are both the same.**

8.6 GAMETE FORMATION AND ALLELES

- During meiosis—gamete formation—alleles that code for the same trait separate into different gametes.

Topic question:

Why did the recessive traits Mendel was studying disappear in the F1 generation? **Because all the offspring were hybrids, meaning each organism in the F1 generation had one recessive and one dominant allele. There were no other combinations of alleles in the F1 generation. No matter which trait Mendel studied, the F1 generation always had one recessive and one dominant allele for the condition. The dominant allele "dominated" the recessive, so none of the F1 organisms displayed any recessive traits.**

8.7 PUNNETT SQUARES

- Punnett squares are used to predict what the possible allele combinations may be when two organisms breed together.

Topic question:

True or False? In order to use a Punnett square, the allele types of both parents must be known. **True. The allele type of one parent is written at the top of the square and the allele type of the other parent is written on the side of the square. The information is then filled in the boxes.**

8.8 F1 CROSS: THE F2 GENERATION

- Mendel crossed the F1 organisms with one another and called the resulting generation the F2 generation.
- Mendel found that some of the F2 plants displayed a re-emergence of the recessive trait that was lost in the F1 generation.

Topic question:

What is the reason that Mendel found the recessive trait reemerged in the F2 organisms? **Since each F1 organism contained a recessive allele, during gamete formation, there were gametes that formed containing only the recessive allele for the trait. If a male parent recessive gamete and a female parent recessive gamete from the F1 generation which each contained a recessive allele combined with one another, then that F2 offspring would have both recessive alleles for the condition and would display the recessive trait.**

8.9 INCOMPLETE DOMINANCE INHERITANCE PATTERN

- Incomplete dominance results when no trait coded for by an allele of a gene is dominant over the allele or alleles.

Topic question:

What happens when a red flower gamete combines with a white flower gamete, and the alleles have an incomplete dominance-type of relationship? **The result is a blend of the traits of the two alleles. For example, when a red flower and a white flower are crossed, the result is a pink flower. The pink is a blend of the red and the white.**

8.10 MULTIPLE GENE INHERITANCE PATTERN

- In multiple gene inheritance, two or more genes affect how a trait is expressed.

Topic question:

What is the term for the condition in which more than one gene affects the way a trait is expressed? **Multiple gene inheritance pattern.**

8.11 SEX-LINKED INHERITANCE PATTERN

- The sex chromosomes do not just carry genes that code for sexual characteristics. They also carry non-sex traits and these are called sex-linked traits.

Topic question:

True or False? A sex-linked trait is carried on a sex chromosome, but does not control the development of a sexual characteristic. **True.**

8.12 SEX-LINKED DISEASES

- When a gene that codes for a sex-linked trait is defective, it can result in a sex-linked disease.
- Males are much more likely to have sex-linked diseases than females.

Topic questions:

Why are males more likely to have a sex-linked disease than females? **The sex-linked diseases are almost always carried on the X chromosome. Females have two X chromosomes, and it is unlikely that a female will inherit a defective sex-linked trait from her mother and her father. If she only has one sex-linked gene that is defective, the normal sex-linked gene on the other X chromosome will counteract it. Males only have one X chromosome, always inherited from the mother. If that one X chromosome has a defective sex-linked gene, there is no normal gene to counteract it on the Y chromosome, so the male is much more likely to have a sex-linked genetic disease.**

True or False? A female is considered a carrier if she has a defective sex-linked gene but does not have the disease for which it codes. **True.**

8.13 GENE-BASED GENETIC DISEASES

- Genetic diseases are caused by defective proteins for which the defective genes code.

Topic question:

True or False? Duchenne muscular dystrophy is caused by a sex-linked gene that codes for the production of a defective muscle protein. **True.**

8.14 AUTOSOMALLY INHERITED GENETIC DISEASES

- Genetic diseases inherited on autosomes are called autosomally inherited genetic diseases.

Topic question:

How many defective alleles does a person need to have in order to have an autosomal genetic disease? **Two—both alleles that code for the same protein need to be defective.**

8.15 MUTATION

- Mutations are the cause of defective genes and the basis of genetic diseases.

Topic question:

True or False? Genes are altered and made abnormal due to infections. **False. Mutations cause defective genes.**

8.16 CHROMOSOME-BASED GENETIC DISEASES

- Some genetic diseases are caused by too many or too few chromosomes being passed from one parent to a child.

Topic question:

Down syndrome is a type of what genetic disease? **A chromosomal genetic disease. A person with Down syndrome has three chromosome number 21 instead of two.**

8.17 DNA TECHNOLOGY

- Genetic engineering/DBA technology is the use of our knowledge of genetics for the advantage of humans.

Topic questions:

True or False? Selective breeding is an example of genetic engineering. **True.**

Can an organism be given a trait it does not have? **Yes, that is the basis of genetic engineering—figuring out a way to insert a gene into something and give it a trait it does not have.**

8.18 KEY CHAPTER POINTS

- Heredity is the passage of traits (genes) from one generation to the next.
- Genetics is the study of heredity.
- Gregor Mendel identified the basic principles of genetics 150 years ago during his study of pea plants.
- Different forms of a gene that code for different traits are called alleles.
- Every trait has two alleles that code for it, one on the chromosome of a pair inherited from the male parent and one on the other chromosome of the pair inherited from the female parent.
- Punnett squares are helpful tools for predicting gamete types based on the alleles of the parents.
- Dominant traits (alleles) suppress the expression of recessive alleles (traits).
- The only way a recessive trait can be expressed is if the organism has two copies of the same recessive allele.
- There are many diseases resulting from defective genes; these are called genetic diseases.
- Genetic engineering is the use of our knowledge of DNA for the betterment of human life.

9 Evolution and Creation

9.0 CHAPTER PREVIEW

In this chapter we will:

- Discuss the two most common beliefs about the origin of life—creation and evolution.

- Emphasize that both evolution and creation are faith-based beliefs regarding the emergence and existence of life.

- Explore the creation and evolution-endorsed process of natural selection.

- Investigate whether neo-Darwinism can properly explain how new types of organisms form.

- Investigate why creationists do not believe evolution adequately explains the existence of life and vice-versa.

- Discuss the process of fossil formation.

- Discuss the ways fossils and radiodating are interpreted by evolutionists and creationists.

- Investigate the possibility of transitional forms in the fossil record; specifically, *Archaeopteryx*.

9.1 INTRODUCTION

- Evolution is the idea that life originated on earth accidentally, the result of pure chance.

- Creation is the belief that life was created by an all-powerful Creator—God—and that life was created specifically for a purpose.

Topic question:

True or False? Evolutionists believe that life is on this planet for a reason and creationists believe life exists here by events that occurred through random chance. **False. Evolution is the belief that life arose on the planet through random events. Creation is the belief that life originated through the specific creation by God and that all life has a purpose.**

9.2 THE TRUTH ABOUT EVOLUTION

- Evolution is nothing more than an atheist's belief of how life came to be here just as creation is a creationist's belief of how life originated.

Topic question:

True or False? Evolution is an atheist's way of describing how life originated on earth. **True.**

9.3 CHARLES DARWIN AND EVOLUTION

- Charles Darwin's book, *The Origin of Species* (1859), is considered the basis for modern evolutionary thought.

- Darwin started life as a believer in God, but ended life as a nonbeliever.

- His voyage on the ship *H.M.S. Beagle* played an important part in forming his ideas.

Topic question:

True or False? Charles Darwin is considered the father of evolution. **True.**

9.4 NATURAL SELECTION

- Natural selection is also termed survival of the fittest.

- Organisms with traits that cause them to be better suited to living in a certain environment are selected for living in that environment and live to pass their traits to their offspring. Organisms that do not have beneficial traits are selected against through natural selection and do not live to pass their non-beneficial traits to their offspring.

- A key evolutionary belief is that natural selection leads to the formation of new types of animals, called species.

- The chance that natural selection could have randomly resulted in the first cell forming is such a small number that creationists do not believe it could happen.

Topic questions:

True or False? Evolutionists believe that the first life-form was single-celled structure, such as a prokaryote, which assembled into a cell randomly. **True.**

True or False? Evolutionists believe the earth is billions of years old. **True.**

9.5 NEO-DARWINISM

- **Neo-Darwinism** is the evolutionary attempt to explain how evolution occurs on a genetic level.

Topic question:

Can organisms evolve if they do not acquire the genes that code for the new traits that cause them to evolve? **No. In order for evolution to cause a less-complicated organism to evolve into a more complicated one, the less-complicated organism needs to acquire new genes that code for the development of the more complicated species' structures and actions.**

Topic question:

How do evolutionists propose that less-complicated organisms acquire new genes? **Through mutations.**

9.6 PHYLOGENETIC TREES

- Phylogenetic trees are tools used by evolutionists to demonstrate relationships between organisms of different types.

Topic question:

True or False? According to the phylogenetic tree, humans are most closely related to gorillas. **False. According to the phylogenetic tree, humans are most closely related to chimpanzees.**

9.7 THE FAITH OF EVOLUTION

- Evolution requires great faith because:

  - There has never been one mutation event identified that adds meaningful, genetic material.

  - Statistical evidence indicates the likelihood of the first cell forming is less than a 1 in $10^{57,000}$ chance.

Topic question:

Evolution is based on hard fact and the basic belief of neo-Darwinism has been proven over and over again. **False. Evolution relies on faith that certain things happened the way evolutionists believe they did just as creation is based on faith that certain things happened the way creationists believe they did.**

9.8 CREATION

- Creationists believe that an all-powerful, all-knowing Creator was able to make the universe out of nothing, resulting in the creation of everything that was intelligently designed.

Topic questions:

What does intelligent design mean? **Since God is all-powerful, all-knowing, and knows what the plans are for everything He creates, He would have created every organism with the exact amount of DNA and genes it needs to fulfill His plan.**

What objections do evolutionists have to creation? **Evolutionists do not believe it is scientific to believe in something that cannot be proven, and they do not believe the existence of God can be proven. Therefore to believe in His creation is not scientific.**

9.9 DIVERSITY

- Diversity is another word for the many different types of organisms on the planet.
- Evolutionists believe diversity exists due to information-adding mutations; creationists believe diversity exists because God created it that way.

Topic question:

True or False? The belief that information adding mutations are an explanation for the diversity of life is based on scientific fact. **False. There has never been one information-adding mutation proven to occur, nor has there ever been one induced in a scientific setting. Evolutionist's have a faith-based belief that information-adding mutations occur since they have never been proven to occur.**

9.10 NATURAL SELECTION

- Creationists and evolutionists agree on this issue—natural selection does occur as described by Blythe and Darwin.

Topic question:

True or False? Creationists do not believe that natural selection occurs. **False. They do believe it occurs.**

9.11 CREATIONISM, NEO-DARWINISM, AND NATURAL SELECTION

- Evolutionists believe that information-adding mutations provide trait differences between organisms that are acted on by natural selection. Over time, natural selection leads to formation of new species.
- Creationists believe that is not true and believe that God created all organisms with great genetic diversity—as much as they would need to fulfill his plan for them.

Topic question:

Is the example of a dark-living species of animal losing their eyes over time evolution? Why or why not? **It is not evolution. Evolution is by definition a less complex organism being transformed into a more complex organism by gene-adding mutations. When the dark-living organisms lost their eyes, they did so through mutation. However, the mutations caused two things: a loss of genetic information (the information coding for the development of eyes), and the movement of a more complex organism into a less complex one (an organism without eyes is obviously less complex than one with eyes). This is not evolution.**

9.12 WHAT IF GENE ADDING MUTATIONS DO OCCUR?

- Even if gene-adding mutations occurred, it seems unlikely that the organisms would have been able to survive the process of natural selection because the intermediate structures formed would make an organism less likely to survive, not more likely.

Topic question:

True or False? If information-adding mutations did occur, they would likely make an organism less likely to survive in their environment and cause them to be killed through natural selection. **True.**

9.13 CREATION, EVOLUTION, AND THE FOSSIL RECORD

- Fossils are often used by evolutionists to prove that the earth is old and one organism transformed into another.

- According to evolutionary theory, the fossil record should be full of transitional forms. It is not. This is viewed as a major flaw of evolutionary thought, even by evolutionists.

Topic questions:

True or False? A transitional form is an animal that shows traits of the organism it is evolving from and changing into. **True.**

True or False? The fossil record is full of transitional forms. **False. There is not one fossil that has been accepted to be a transitional form.**

9.14 SUMMARY

- It seems only fair to accept evolution and creation for what they are. Evolution is a faith-based philosophy that atheists devised to explain the origin of life and species in order to remove God from the equation. Creation is a faith-based philosophy that God created everything.

Topic question:

True or False? The reason that evolution is so appealing as an explanation for the origin of life and species is because it is supported by hard scientific facts. **False. Evolution requires as much faith to believe as creation.**

9.15 KEY CHAPTER POINTS

- Evolution is the faith-based belief system that life originated on earth randomly, and that less-complex life forms evolved into more-complex life forms.

- Creation is the faith-based belief system that God created the universe and everything in it for His purpose.

- Natural selection is a biological process in which organisms with beneficial traits are able to live and reproduce, passing their beneficial traits on to their offspring, while the organisms with less beneficial traits do not reproduce, causing their non-beneficial traits to die off with them. Creationists and evolutionists identify this observed process as being true.

- Neo-Darwinism is the evolutionary attempt to explain evolution on a genetic basis. There is no scientific basis for the belief in neo-Darwinism; believing in it is faith-based.

- Creationists believe that all the species were created by God and given built-in genetic diversity, so they could adapt and respond to natural selection.

- According to evolutionary thought, there should be transitional forms in the fossil record; there are none.

- According to evolutionary thought, fossils form slowly over time. There is ample evidence that fossils can form quickly. In fact, decomposition of the organism would prevent it from ever being fossilized if fossils do form slowly.

10 Scientific Classification I:
Overview, Archaebacteria, Eubacteria, and Viruses

10.0 CHAPTER PREVIEW

In this chapter we will:

- Introduce the concepts of classifying all living organisms into the seven-level classification system.
- Define taxonomy and the six kingdoms in the seven-level classification system.
- Define what a binomial name is.
- Learn the use of a dichotomous key.
- Investigate the properties of the organisms classified into Archaebacteria and Eubacteria.
- Discuss the properties of the particles called viruses

10.1 OVERVIEW

- All organisms are classified using a seven-level, six-kingdom system.

Topic question:

True or False? A kingdom is the largest level of categorization. **True.**

10.2 TAXONOMY

- Taxonomy is the science of classifying organisms. Scientists who specialize in classifying organisms are called taxonomists.
- Organisms are systematically classified into the seven levels with each receiving a unique binomial name.

Topic question:

Why is a binomial name helpful in the seven-level system? **Since there is only one species that has a given binomial name, there is no confusion when scientists describe the organisms they are studying.**

10.3 DICHOTOMOUS KEYS

- Dichotomous keys are tools to help classify organisms based on their shared and differing characteristics. It helps to identify organisms by asking paired questions regarding the organism's traits, then giving instructions based on the answer.

Topic question:

Are dichotomous keys helpful for a person to identify types of bacteria as well as animals? **Yes. They are helpful to identify any type of living organism.**

10.4 KINGDOMS ARCHAEBACTERIA AND EUBACTERIA

- All prokaryotes are classified into the Kingdoms of Archaebacteria or Eubacteria.
- Archaebacteria often live in hostile environments.
- Eubacteria are the typical kinds of bacteria that cause infections. They have cell walls that are different in structure from the Archaebacteria.

Topic questions:

What are the three cell shapes that bacteria display? **Spiral, coccus (spherical or round) and bacillus (rod).**

What are peptidoglycans? **They are the molecules that make up the bacterial cell wall.**

10.5 VIRUSES

- Viruses are usually not classified in the seven-kingdom system because they are not considered "alive." Viruses are considered particles.

Topic questions:

What is a virus considered if it is not a living organism? **A particle**.

What is a host and why does a virus need one? **A host is the cell or organism a virus infects. Viruses need hosts so they can take over the protein-making machinery of the host cell and make more of the virus.**

10.6 KEY CHAPTER POINTS

- All living organisms are classified into a seven-level system of classification.
- The present classification system is a standard that is used by biologists all over the world.
- Carolus Linnaeus devised the current classification system in the 1700s.
- Using Linnaeus's system, all organisms are given a binomial name, and only one specific organism has a given binomial name.
- A dichotomous key is a simple tool that can be used to classify organisms based on the traits they share and those in which they differ.
- The kingdoms Archaebacteria and Eubacteria contain all the prokaryotic cells, called "bacteria."
- Bacteria are important decomposers.
- Eubacteria are the most numerous organisms on earth and have cell walls composed of peptidoglycans.
- Archaebacteria live in extreme environments and have cell walls that are made of different molecules than Eubacteria.
- Viruses are called particles.

11 Scientific Classification II:
Protista and Fungi

11.0 CHAPTER PREVIEW

In this chapter we will:

- Introduce the basic groups that make up Protista.
- Investigate the properties of the following protist groups:
  - the plant-like protists, or algae
  - the animal-like protists, or protozoa
  - the fungus-like protists, or slime molds
- Introduce the basic structure and functions of Fungi.
- Review the important history of Fungi as it relates to the discovery of antibiotics.
- Discuss the structure and function of mushrooms and molds.
- Define mycorrhizae and their importance to plants and fungi.
- Study the properties of lichens.

11.1 INTRODUCTION TO KINGDOM PROTISTA

- Organisms classified in this kingdom are called protists.
- There are three types of protists—animal-like protists (protozoa), plant-like protists (algae), and fungus-like protists (slime molds).

Topic question:

- What are three types of protists? **Animal-like protists (protozoa), plant-like protists (algae), and fungus-like protists (slime molds).**

11.2 PROTISTA: PROTOZOANS

- Protozoans are called the animal-like protists because, like animals, protozoans need to eat other organisms to obtain their organic molecules and energy.

Topic question:

True or False? Amoeba and paramecium are common types of protozoans. **True.**

11.3 PROTISTA: ALGAE

- Algae are called the plant-like protists because most of them are green, contain chloroplasts to perform photosynthesis, and have cell walls.

Topic question:

True or False? The plant-like protists, algae, do not perform photosynthesis. **False. Algae contain chloroplasts and chlorophyll and perform photosynthesis.**

11.4 PROTISTA: SLIME MOLDS

- Slime molds, like fungi, obtain their energy and organic molecules by decomposing dead organic material.

Topic question:

Why are slime molds called fungus-like protists? **They both obtain their energy the same way, through the digestion of dead material (they are both decomposers).**

11.5 PROTISTA: REPRODUCTION

- Protists can reproduce both sexually and asexually.

Topic question:

What is the difference between conjugation and fragmentation? **They are performed by different protist species. Fragmentation is performed by algae and conjugation by protozoa. Also, fragmentation is asexual reproduction and conjugation is sexual reproduction.**

11.6 KINGDOM FUNGI

- Organisms in Fungi are characterized by cell walls made from chitin and by being saprophytes.
- Hyphae are the basic structural unit of all Fungi.

Topic questions:

What is a saprophyte? **It is an organism that obtains its nutrition by eating dead organisms. Decomposers are saprophytes.**

What is a hyphus? **It is the basic structural unit of fungi and is composed of many hyphal cells linked to one another end to end.**

11.7 FUNGI: HISTORY

- Some fungi are helpful to man, such as *Penicillium*, where we obtained penicillin, and yeast, for leavening bread.

Topic question:

True or False? Fungi can be both helpful and harmful to man. **True.**

11.8 FUNGI: MUSHROOMS

- Mushrooms are a common fungus.
- They have a common structure of a stalk and cap. Their body is called a mycelium.

Topic question:

True or False? The mycelia of mushrooms that sit on top of their food secrete enzymes to digest the food, then the mycelia absorb the released nutrition. **True.**

11.9 FUNGI: MOLDS

- Molds have a common structure. The hyphae that extend into their food source are called rhizoids. Other hyphae grow across the top of their food, called stolons. Those that extend up off the food surface and form spores are called sporangium.

Topic question:

True or False? Stolons are hyphae that grow down into the food source. **False. Stolons are hyphae that extend across the surface of the food source.**

11.10 FUNGI: REPRODUCTION

- Fungi can reproduce sexually and asexually.
- Spores are the asexual reproductive structures of Fungi.

Topic question:

True or False? Asexual reproduction of Fungi occurs when hyphae from different fungi of the same species fuse and form a new organism. **False. That is sexual reproduction in Fungi. Asexual reproduction is through spores.**

11.11 MYCORRHIZAE

- Mycorrhizae is a term used to describe the symbiotic relationship most plants share with fungi.

Topic question: see next section

11.12 LICHENS

- Lichens are a symbiotic relationship between Fungi and algae.

Topic questions:

Lichens are a symbiotic relationship of Fungi with what other organism? **Algae.**

True or False? Fungi can form symbiotic relationships with plants and algae. **True.**

11.13 KEY CHAPTER POINTS

- Protista is made up mainly of unicellular organisms, although there are some multicellular forms.
- The three groups of protists are the animal-like protists (protozoa), the plant-like protists (algae), and the fungus-like protists (slime molds).
- Protists form a large component of plankton.
- Protozoa do not have cell walls and need to eat other organisms to obtain their organic molecules and energy.
- Protozoa can move in three different ways: flagella, cilia, and amoeboid movement.
- Algae are photosynthetic, contain chloroplasts, and have cell walls.
- Slime molds are decomposers.
- Protists can sexually (usually) and asexually (less commonly) reproduce.
- Fungi are decomposers with cell walls made of chitin.
- The basic unit of a fungi's structure is called a hyphus. Many hyphae form the structure of the fungus organism.
- Fungi can reproduce sexually or asexually.
- Mycorrhizae are the symbiotic relationship that fungi have with plants.
- Lichens are the symbiotic relationship Fungi have with many algal species.

12 Scientific Classification III:
Plantae I

12.0 CHAPTER PREVIEW

In this chapter we will:

- Investigate Plantae classification using a dichotomous key.

- Define and discuss the following plant traits:
  - vascular tissue
  - spore producing or seed producing
  - gymnosperm
  - angiosperm
  - monocot
  - dicot

- Discuss the characteristics and examples of:
  - nonvascular, seedless plants
  - vascular, seedless plants
  - gymnosperms
  - angiosperms

- Further divide the angiosperms into monocots and dicots.

- Investigate root, stem, and leaf structures and functions.

12.1 INTRODUCTION

- Scientists who study plants are called **botanists**. The study of plants is called **botany**. Plants are classified in the Kingdom **Plantae**.

- All plants have cell walls made of cellulose, chloroplasts to perform photosynthesis, and are eukaryotes.

Topic question:

What three things do all plants have in common? **They have cell walls made of cellulose, chloroplasts for photosynthesis, and are eukaryotes. Also, I would accept the answer of they do not move, but that is "iffy."**

12.2 PLANT STRUCTURE AND CLASSIFICATION

- The plant cell structure is reviewed.

Topic questions: none

12.3 DICHOTOMOUS KEY FOR PLANTAE

- Like all other organisms, there are many dichotomous keys in use to classify plants.

- Plants are generally classified into: nonvascular; vascular and non-seed-producing; vascular, seed-producing, and non-flowering; and vascular, seed-producing, and flowering.

Topic question:

True or False? There are four general categories of plants. **True.**

12.4 VASCULAR AND NONVASCULAR PLANTS

- The vascular tissue of plants is a series of connected tubes that run from the tips of the roots to the tips of the leaves and everywhere in between. Plants with this type of plumbing are called vascular plants.
- Plants that do not have this type of plumbing are called nonvascular plants.
- The majority of plants have vascular tissue.

Topic question:

Are the majority of plants vascular or nonvascular? **Vascular.**

12.5 NONVASCULAR: MOSSES, LIVERWORTS, AND HORNWORTS

- Mosses, liverworts, and hornworts are some of the common names of the nonvascular plants.
- Osmosis moves water through the plant from cell to cell. Diffusion moves nutrients and minerals from cell to cell.

Topic questions:

How do nonvascular plants transport substances through the plant? **Osmosis and diffusion.**

True or False? The sexually-reproducing, nonvascular moss plant grows out of the non-sexually-reproducing moss plant. **True.**

12.6 VASCULAR PLANTS

- Plants have specialized tissues that form tubes all through the plant.
- Xylem is the vascular tissue that carries water and minerals from the roots to the rest of the plant.
- Phloem carries glucose from the leaves (where photosynthesis occurs) to the rest of the plant.

Topic questions:

True or False? Xylem and phloem are continuous tubes found only in the leaves and roots. **False. Vascular tissue runs throughout the plant.**

True or False? There are two basic types of vascular plants, those that form seeds and those that do not. **True.**

12.7 SEEDLESS VASCULAR PLANTS

- The seedless vascular plants are plants that have xylem and phloem but reproduce with spores; they do not form seeds.
- The seedless vascular plants are the club mosses, ferns, and horse tails.

Topic questions:

True or False? Ferns and horsetails reproduce with spores. **True.**

True or False? Ferns only have one form of the plant, in which the plant reproduces only asexually with spores. **False. Ferns have two plant cycles (or plant forms)—a cycle (or form) in which the fern plant reproduces sexually and another cycle (or form) in which the fern plant reproduces asexually with spores**.

12.8 SEEDS

- Seeds and spores are structurally different.
- A seed is larger than a spore and has a supply of food surrounding the embryo plant which is, in turn, surrounded by a tough coating.

Topic question:

What is endosperm? **It is the supply of food contained in a seed.**

12.9 FLOWERS

- Not all plants produce flowers.
- Vascular plants that produce seeds but do not make flowers are gymnosperms, also called conifers and pine trees.
- Vascular plants that produce both seeds and flowers are called angiosperms.

Topic question:

What is a plant that has xylem and phloem and produces seeds but not flowers called? **A gymnosperm (conifer and pine tree are okay answers, too).**

12.10 NON-FLOWERING SEED VASCULAR PLANTS: GYMNOSPERMS

- Gymnosperms produce cones to protect their seeds. A pine cone is the reproductive structure of a gymnosperm.
- Most gymnosperms have leaves that look like needles.

Topic question:

True or False? A cone is the asexual reproductive structure of a gymnosperm. **False. It is the sexual reproductive structure.**

12.11 FLOWERING SEED VASCULAR PLANTS: ANGIOSPERMS

- Angiosperms are also called the flowering plants.

Topic question:

True or False? Angiosperms are the second most numerous plant species, behind the gymnosperms. **False. They are the most numerous.**

12.12 MONOCOTS AND DICOTS

- Angiosperms are divided into two groups based on the structure of their seed.
- Monocots produce seeds with one cotyledon.
- Dicots produce seeds with two cotyledons.

Topic question:

What is the difference between a monocot and a dicot? **A monocot plant produces seeds with one cotyledon and a dicot produces seeds with two cotyledons.**

12.13 PLANT STRUCTURE: ROOTS

- Monocot root systems usually form many roots of the same size. This is called a fibrous root system.
- Dicot roots form one main root with several small roots branching from the main root. This is called a taproot system.

Topic question:

What is the difference between monocots and dicots as far as root structure is concerned? **Monocots usually have a fibrous root system and dicots usually have a taproot system.**

12.14 STEMS

- There are two types of stems—herbaceous and woody.
- Monocot stems have their xylem and phloem scattered throughout the stem.
- The xylem and phloem in dicots is arranged in a ring around the center of the stem.
- Tree rings form due to the circular arrangement of xylem and phloem around the center of the tree.

Topic questions:

What is the difference between monocot and dicot stems as far as their vascular bundles (xylem and phloem) are concerned? **Monocot stems have their xylem and phloem scattered throughout the stem. The xylem and phloem in dicots is arranged in a ring around the center of the stem.**

True or False? Tree rings form because the tree grows faster in the late summer and fall than in the spring and early summer, which causes the xylem to form and look different. **False. They form because the tree grows faster in the spring and early summer than in the late summer and fall.**

12.15 MERISTEMS

- Plants grow from tissue called the meristem.

Topic question:

What is the difference between the apical and lateral meristems? **Apical meristems are found at the tips of stems and roots. When apical meristem growth occurs, the roots and stems get longer. Meristems are also found between the xylem and phloem. This is called lateral meristem tissue. Growth of the lateral meristem tissue causes plant parts to get larger in girth ("bigger around"). Lateral meristem tissue growth also forms new xylem and phloem.**

12.16 LEAVES

- Leaves are the main photosynthetic tissue of plants.
- Leaves are designed to limit water loss.
- Leaf veins are the xylem and phloem running through the leaf.
- Leaves from a monocot have veins running along the length of the leaf blade, parallel to one another.
- Dicot leaf veins look like a net. There is one larger vein running down the center of the leaf, with smaller veins branching off the main one.

Topic question:

What is the difference between monocot and dicot leaves regarding the vascular formation? **Monocot vasculature runs parallel, dicot vasculature intersects in a branching pattern.**

12.17 KEY CHAPTER POINTS

- All plants have common characteristics, including eukaryotic cell structure, cell walls made of cellulose, and photosynthesis.
- Like all organisms, plants can be classified based on common and differing traits.
- Traits used to classify plants include:
  - vascular/nonvascular
  - seed production/no seed production
  - flowering plants/non-flowering plants
  - monocot/dicot seed structure
- Nonvascular plants do not produce seeds and are the hornworts, liverworts, and mosses.
- Vascular, non-seed-producing plants are the horsetails and ferns.
- Vascular, seed-producing, non-flowering plants are the gymnosperms.
- Vascular, seed-producing, flowering plants are the angiosperms.
- Angiosperms that produce seeds with one cotyledon are monocots.
- Angiosperms that produce seeds with two cotyledons are dicots.
- Stems lift leaves off the ground and are for transport.
- Roots come in three types—taproot, fibrous, and adventitious.
- Plant growth occurs in the meristem tissue.
- Leaves perform photosynthesis and have stoma to limit water loss.

Scientific Classification IV:

13 | Plantae II

13.0 CHAPTER PREVIEW

In this chapter we will:

- Discuss methods of asexual plant reproduction.
- Discuss the two phases of the plant life cycle—the sporophyte and gametophyte phase.
- Discuss specific reproductive cycles for nonvascular plants, seedless vascular plants, gymnosperms, and angiosperms.
- Investigate flower structure and how it relates to angiosperm reproduction.
- Discuss the process of pollination, fertilization, and germination.
- Investigate the specific roles water has in maintaining plant life.
- Study nastic and tropic plant movements and how they occur.
- Identify the processes responsible for transporting materials in xylem.
- Discuss photoperiodism.

13.1 INTRODUCTION

- This chapter will focus on the reproductive cycles of various groups of plants as well as the importance of water in plant's life.
- Plants can sexually and asexually reproduce.

Topic question: none

13.2 ASEXUAL PLANT REPRODUCTION

- There are a number of ways plants asexually reproduce.

Topic question:

True or False? Grafting, stolons, and stem cutting are all ways that plants can asexually reproduce. **True.**

13.3 SEXUAL PLANT REPRODUCTION: GENERAL

- Plants exhibit two phases of their reproductive cycles: a spore producing sporophyte phase and a gamete-producing gametophyte phase.
- The sporophyte and gametophyte phases are actual plants. Instead of being called a "plant," they are called "a gametophyte" and "a sporophyte."

Topic question:

Which form, or phase, of a plant are gametes produced? **The gametophyte phase.**

13.4 PLANT REPRODUCTION: NONVASCULAR PLANTS

- During the sporophyte phase, nonvascular plants form spores.
- Spores germinate and grow into gametophytes, which grow right out of the sporophyte plant.
- The gametophyte makes gametes—sperm and eggs. The sperm fertilizes the egg and a zygote is formed.
- The zygote grows into a sporophyte plant and the cycle starts over.

Topic questions:

How is a nonvascular plant sporophyte formed? **The gametophyte plant produces sperm and eggs. A sperm fertilizes an egg and forms a zygote. The zygote then grows into the sporophyte plant.**

True or False? The gametophyte form of the nonvascular plants usually grows from the top of the sporophyte form of the plant. **True.**

13.5 PLANT REPRODUCTION: VASCULAR SEEDLESS PLANTS

- The vascular seedless plants follow the alternation between gametophytes and sporophytes from one generation to the next.
- The phase of the plant that is recognizable as a fern is the sporophyte phase.

Topic questions:

True or False? The large fern plant that is recognizable as a fern is a fern in the sporophyte phase. **True.**

Where do fern spores form? **On the undersurface of the leaves (fronds).**

13.6 PLANT REPRODUCTION: GYMNOSPERMS

- Pine trees (conifers) have male and female cones, which produce male and female spores.
- "Pine trees" are actually gymnosperms in the sporophyte phase.
- The male spore grows into a male gametophyte plant and the female spore grows into a female gametophyte plant.
- The male gametophyte plant produces the male gamete, sperm, which fertilizes the female gametophyte, the egg.
- This forms a zygote, which is enclosed in a seed.
- When the seed grows into a new plant, it will be the sporophyte phase.

Topic questions:

True or False? A conifer (gymnosperm) male gamete fertilizes the female gamete in the pine cone. **True.**

True or False? A conifer spore grows into a gametophyte. **True.**

True or False? Male and female conifer (gymnosperm) gametophytes are large structures. **False. They are small structures that grow in pine cones.**

13.7 PLANT REPRODUCTION: ANGIOSPERMS AND FLOWERS

- Flowers are the reproductive structures of angiosperms.
- The structure of a flower is often dependent on whether an animal/insect or the wind pollinates it.
- Stamens are the male reproductive structure; they make pollen.
- Pistils are the female reproductive structure; they form eggs in the ovary.

Topic questions:

True or False? Plants that are pollinated by the wind often contain nectar. **False. Plants that are pollinated by the wind usually do not have nectar. Plants that are pollinated by insects/animals usually have nectar.**

What is the female reproductive structure of a flower called? **The pistil.**

What is the male reproductive part of a flower called? **The stamen.**

13.8 PLANT REPRODUCTION: ANGIOSPERMS AND FERTILIZATION

- Fertilization is the process of the sperm fusing with the egg.
- A tube quickly forms from the pollen grain. The tube grows from the pollen through the style to the ovary. A sperm cell moves through this passage to the egg in the ovule. Fertilization occurs, and a seed is formed.

Topic question:

How does fertilization occur in angiosperms (following pollination). **A tube quickly forms from the pollen grain. The tube grows from the pollen through the style to the ovary. A sperm cell moves through this passage to the egg in the ovule. The sperm unites with the egg and the chromosomes from the male parent are combined with those of the female parent. A zygote is formed and is covered by a seed coat.**

13.9 GERMINATION

- **Germination** is the process of a seed sprouting and beginning to grow into a new plant.
- In order for a seed to sprout and grow, the right conditions need to be met.

Topic question:

What is germination? **The process of a seed sprouting and beginning to grow.**

13.10 WATER: PHOTOSYNTHESIS

- Water is critical for photosynthesis to occur.

Topic question:

Describe the function of guard cells. **The guard cells open and close the stoma on the undersurface of the leaf depending on water conditions. If water is in short supply, the guard cells close the stoma so photosynthesis cannot occur. This saves water so the plant does not dry out. When water is plentiful, the guard cells open the stoma again. By doing so, carbon dioxide can enter the leaf so photosynthesis can occur.**

13.11 WATER: TURGOR PRESSURE

- Turgor pressure is pressure that builds up inside of a cell due to water.

Topic question:

What lost property of a plant is responsible for wilting? **Turgor pressure.**

13.12 WATER: NASTIC MOVEMENTS

- Nastic movements of a plant are physical movements in a plant that occur repeatedly throughout the day or the plant's entire life and are the result of changes in turgor pressure.

Topic question:

How do nastic movements occur? **Due to changes in turgor pressure.**

13.13 WATER: TRANSPORTATION

- Plants are able to move substances through their xylem because of transpiration.

Topic question:

True or False? Transpiration provides a "pulling" force on water inside of xylem and serves to "pull" water upward in the xylem. **True.**

13.14 PLANT HORMONES

- Hormones are chemicals that are produced in one area of an organism and affect the growth or chemical reactions in another area of the organism.
- There are several different types of plant hormones—auxin, gibberellin, ethylene, cytokinins, and abscisic acid.

Topic questions:

What do plant hormones usually regulate? **Plant cell growth.** What does ethylene do to a fruit? **It causes it to ripen quickly.**

13.15 TROPISMS

- Tropisms are the directional movements of a plant due to a stimulus; they are regulated by hormones.

Topic question:

How are tropism and nastic movements different? **Tropisms occur due to the effects of hormones; nastic movements occur due to the effects of turgor pressure.**

13.16 PHOTOPERIODISM

- Photoperiodism relates to a certain process of a plant responding to the amount of light it receives.

Topic question:

True or False? The time of year that many plants produce flowers is dependent on the amount of light they receive, which is an example photoperiodism. **True.**

13.17 KEY CHAPTER POINTS

- All plants are capable of asexual reproduction.
- All plants show two separate phases during their reproductive cycles. One phase is a spore-producing (sporophyte) phase, the other is a gamete-producing (gametophyte) phase.
- Spores are small cells capable of developing and growing into a gametophyte.
- The male gametes (sperm) need to fuse with a female gamete (egg) to form a new organism. This new organism grows into a sporophyte.
- Flowers are specialized angiosperm reproductive structures.
- The male reproductive structure of a flower is the stamen; the female reproductive structure is the pistil.
- Pollination is the process of transferring pollen from the anther to the style. This can occur by wind or pollinators.
- Fertilization is the process of the male gamete fusing with the female gamete and combining their chromosomes to form a new organism.
- Germination is the process of a seed embryo breaking through the seed coat and starting to grow.
- Water is responsible for maintaining turgor pressure and for xylem flow.
- Plants produce hormones that control various processes and growth.
- There are two types of plant movements—nastic and tropisms. Water is responsible for nastic movements and hormones for tropisms.

Scientific Classification V:

14 | Kingdom Animalia I

14.0 CHAPTER PREVIEW

In this chapter we will:

- Introduce technical terms related to the classification of animal species.

- Prepare a dichotomous key to classify all animal phyla discussed in this course.

- Discuss the features of the following phyla:
  - Porifera
  - Cnidaria
  - Platyhelminthes
  - Annelida
  - Mollusca
  - Arthropoda
  - Echinodermata

14.1 INTRODUCTION

- All animals are classified into the Kingdom Animalia.

- Biologists who study animals are called zoologists.

Topic question:

How do zoologists classify animals? **Zoologists use similarities and differences in an animal's structure and function to classify them.**

14.2 ANIMAL PROPERTIES

- Almost all organisms in Animalia share the following characteristics:
  - eukaryotic cells
  - lack of cell walls
  - multicellular structure
  - extra-cellular matrix
  - cells that are organized into tissue
  - sexual reproduction
  - mobility
  - heterotrophic
  - the ability to store excess energy as glycogen
  - aerobic

Topic question:

True or False? Animals have the following properties in common: no cell walls, extracellular matrix, nucleoid, and sexual reproduction. **False. The nucleoid is not an animal feature.**

14.3 ANIMAL TYPES

- There are two general types of animals—those with spinal columns or vertebrae and those without. Animals with a spinal column are called vertebrates. Those without are called invertebrates.

- A dichotomous key will be used to help the student understand the thought process that goes into classifying animals.

Topic question:

What is the difference between a vertebrate and invertebrate? **A vertebrate has a spinal column, composed of bones called vertebrae, that protects a spinal cord. An invertebrate does not have a spinal column.**

14.4 SYMMETRY

- Animals display the property of symmetry, which is when one half of an animal looks like the other half.

Topic question:

What is the difference between radial symmetry and bilateral symmetry? **Radial symmetry is when the top half of an animal looks like the bottom half. It is a type of symmetry found only in aquatic animals. Bilateral symmetry is when the right half of an animal looks like the left half.**

14.5 THE GUT

- Almost all animals have an internal structure, referred to as a gut, to process their food.

Topic question:

What is a gut? **It is an internal structure that digests and processes food.**

14.6 INVERTEBRATES: PORIFERA

- These organisms are commonly known as sponges.
- Their bodies are filled with holes, called pores.
- They are filter feeders.

Topic question:

Which one of these is not a feature of the organism of Porifera? Sessile, filter feeders, body full of pores, bilaterally symmetric. **Sponges display all those features except for bilateral symmetric.**

14.7 INVERTEBRATES: PHYLUM CNIDARIA

- Cnidarians are better known as jelly fish, coral, hydras, and sea anemones.
- Their defining feature is an organelle, called a nematocyst, not found in any other species.

Topic question:

What is a cnidocyte and what does it contain? **It is a specialized stinging cell found only in cnidarians. It contains a special organelle called the nematocyst, which stings.**

14.8 INVERTEBRATES: PLATYHELMINTHES

- Organisms in this phylum are commonly known as flat worms.
- Included in this phylum are important animal and human parasites.

Topic question:

True or False? Some of the organisms from Platyhelminthes are parasites and others are able to live on their own (free-living). **True.**

14.9 INVERTEBRATES: ANNELIDA

- Organism in this phylum are commonly called earthworms.

Topic questions:

True or False? Earthworms have a cerebral ganglion (worm brain) that is connected to a nerve cord. **True.**

What type of digestive tract do annelids have, one way or two way? **Two way.**

14.10 INVERTEBRATES: MOLLUSCA

- Organisms in this phylum are commonly called mollusks and include octopus, squid, clams, and snails.

Topic question:

True or False? Almost all mollusks have a closed circulatory system. **False. Almost all mollusks have an open circulatory system.**

14.11 INVERTEBRATES: ARTHROPODA

- Animals in this phylum include insects, spiders, lobsters, and trilobites.

Topic questions:

What molecule is an insect's exoskeleton made from? **Chitin.**

Why do insects molt? **Their exoskeletons are rigid and do not allow for growth. As the insect grows, it sheds its exoskeleton so it can continue to grow.**

What are the three parts that divide an insect's body? **Head, thorax, and abdomen.**

Metamorphosis that includes a larval and pupa stage is an example of complete or incomplete metamorphosis? **Complete.**

14.12 INVERTEBRATES: ECHINODERMATA

- This phylum includes the sea stars (star fish), sand dollars, sea urchins, and sea cucumbers.

Topic question:

What is a water vascular system? **It is a series of connected tubes throughout an echinoderm's body that allow water to move through them.**

14.13 KEY CHAPTER POINTS

- Almost all organisms in Animalia share the following characteristics:
    - eukaryotic cells
    - lack of cell walls
    - multicellular structure
    - extra-cellular matrix
    - cells that are organized into tissue
    - sexual reproduction
    - mobility
    - heterotrophic
    - the ability to store excess energy as glycogen
    - aerobic
- Vertebrates have a spinal column and invertebrates do not.
- Animals display one of three types of symmetry: radial, bilateral, or no symmetry.
- Sponges are in the phylum Porifera. Their defining feature is a porous body.
- Jellyfish, anemones, and coral are in the phylum Cnidaria. Their defining feature is a special cell (a cnidocyte), which contains a special organelle (the nematocyst).
- Flatworms are in the phylum Platyhelminthes. Their defining feature is a flat body with no appendages.
- Earthworms are in the phylum Annelida. Their defining features are the many visible body segments and lack of appendages.
- Crabs, octopi, squids, clams, and snails are in the phylum Mollusca. Their defining feature is a soft body.
- Spiders, centipedes, millipedes, insects, lobsters, crabs, and shrimp are in the phylum Arthropoda. Their defining features are an exoskeleton made of chitin and jointed appendages.
- Starfish, sand dollars, and sea urchins are in the phylum Echinodermata. Their defining feature is an endoskeleton made of ossicles.

15 Scientific Classification VI:
Kingdom Animalia II

15.0 CHAPTER PREVIEW

In this chapter we will:

- Discuss the features of the following Chordata classes:
  - Chondrichthyes, the cartilage fish
  - Osteichthyes, the boney fish
  - Amphibia, the amphibians
  - Reptilia, the reptiles

- Discuss the terms endothermic and ectothermic, then apply them to the chordate classes.

- Investigate the function of capillaries.

- Learn about one-loop and two-loop circulatory systems.

15.1 INTRODUCTION

- All vertebrates are members of the phylum Chordata.

Topic question:

True or False? All members of Chordata have an external skeleton. **False. All members of Chordata have an endoskeleton.**

True or False? Some organisms in Chordata have a skeleton made of bone and others have one made of cartilage. **True.**

15.2 ENDOTHERMIA AND ECTOTHERMIA

- Animals that produce their own body heat are called endotherms, also referred to as "warm-blooded." Birds, mammals, and humans are endotherms.

- Cold-blooded animals do not perform as many endothermic reactions as endothermic animals. Cold-blooded animals are called ectotherms.

Topic question:

What is the difference between ectothermia and endothermia? **Endothermic animals perform enough chemical reactions that they are able to generate their own body heat and maintain a constant temperature independently. Ectothermic animals are not able to do this and must rely on the temperature of their environment to maintain a stable body temperature.**

15.3 GAS EXCHANGE

- Gas exchange has two components. One component occurs in the lungs or gills and is the process of bringing oxygen into the body from the atmosphere and releasing carbon dioxide into the atmosphere. The other component occurs in the tissues and is the process of bringing oxygen into the cell from the blood and releasing carbon dioxide from the cell into the blood.

Topic questions:

In which type of blood vessel does gas exchange occur? **Capillaries.**

True or False? Gas exchange can occur in tissues, gills, or lungs. **True.**

15.4 CIRCULATION

- All vertebrates have a closed circulatory system.
- Blood flows form the heart to the tissues in arteries and from the tissues to the heart in veins.
- Some members of Chordata have a one-loop system and others have a two-loop system.

Topic questions:

Describe the blood flow in a one-loop circulatory system. **One-loop systems are only found in aquatic animals. The blood is pumped from the heart to the gills, then continues to move from the gills to the body tissues. It continues to flow back to the heart and completes the single loop.**

True or False? Amphibians have a three-chambered heart. **True.**

15.5 FISH

- There are two types of fish—fish with bones and fish with cartilage.

Topic question:

What is a lateral line? **The lateral line is a group of nerve cells running down both sides of the organism; they collect information from the environment such as vibration and electrical currents. This information is carried to the brain through the dorsal spinal cord. This is a sensitive part of the nervous system, which allows the organism to sense and respond to changes in its environment.**

15.6 AMPHIBIANS

- This group of animals includes frogs, toads, and salamanders.

Topic questions:

Describe the metamorphosis of a frog. **The larval form of a frog is called a tadpole. Frog tadpoles live completely in the water. During the metamorphosis, the tadpole loses its tail and gills, growing legs and lungs. In addition, the circulatory system changes from a one-loop system to a two-loop system. The tadpole heart changes from a two-chambered heart into a three-chambered frog heart.**

True or False? Amphibians lay hard-shelled eggs that hold up to tough conditions. **False. Amphibians lay soft eggs that dry out easily.**

15.7 REPTILES

- This group includes snakes, turtles, lizards, crocodiles, alligators, tuataras, and dinosaurs.

Topic question:

What is the structure and the importance of the amniotic egg? **It is a hard-shelled container that protects the developing embryo. The amniotic egg contains food for the developing organism and disposes of wastes. It decreases the dependence on water that amphibians have for their eggs to survive.**

15.8 KEY CHAPTER POINTS

- Vertebrates have an endoskeleton. The endoskeleton of Chondrichthyes is made of cartilage. Osteichthyes, Amphibia, Reptilia, Aves, and Mammalia have endoskeletons made of bone.
- The tissues of Chordata are organized into complex organ systems.
- Organisms in Chondrichthyes, Osteichthyes, Amphibia, and Reptilia are ectothermic. Aves and Mammalia are endothermic.
- Gas exchange is the process of exchanging oxygen for carbon dioxide in the tissues and carbon dioxide for oxygen in the lungs.
- Gas, nutrient, and waste exchange occurs in capillaries.

- The heart pumps blood in organisms from Chordata.
- Arteries carry blood from the heart; veins carry it to the heart.
- Fish have a one-loop circulatory system.
- Terrestrial vertebrates and aquatic mammals have a two-loop circulatory system.
- Cartilage fish are classified in the class Chondrichthyes.
- Boney fish are classified in the class Osteichthyes.
- Frogs, toads, and salamanders are classified in the class Amphibia.
- Dinosaurs, snakes, lizards, turtles, tuatara, crocodiles, and alligators are classified in the class Reptilia.

16 Scientific Classification VII:
Kingdom Animalia III

16.0 CHAPTER PREVIEW

In this chapter we will:

- Discuss the features of Aves.

- Discuss the general features of mammals.

- Further categorize the mammals into placental mammals, marsupials, and monotremes.

- Further categorize the placental mammals into the following orders:
  - Rodents (Rodentia)
  - Insect-eating mammals (Insectivora)
  - Toothless mammals (Edentata)
  - Flying mammals (Chiroptera)
  - Hoofed mammals (Artiodactyla and Perissodactyla)
  - Trunk-nosed mammals (Proboscidea)
  - Carnivores (Carnivora)
  - Aquatic mammals (Cetacea and Sirenia)
  - Primates (monkeys, apes, and humans)

16.1 INTRODUCTION

- This chapter we will discuss the only two groups of organisms that are endothermic—birds and mammals.

Topic question:

What are the only two groups of endothermic organisms? **Mammals (Mammalia) and birds (Aves).**

16.2 BIRDS

- All birds have beaks, feathers, a light-weight body, and air sacs. They also all lay eggs and are endothermic.

Topic questions:

True or False? Air sacs help to make birds lighter and allow them to perform gas exchange continuously on inhalation and exhalation. **True.**

What types of eggs do birds lay? **Amniotic eggs.**

True or False? There are three general types of birds—perching, flightless, and water birds. **False. There are four. The fourth category is birds of prey.**

16.3 MAMMALS

- All mammals have mammary glands, hair, large brains, a diaphragm, and a single lower jaw. They also all care for their young and have other shared traits.

- There are three groups of mammals—monotremes, marsupials, and placental mammals.

Topic questions:

True or False? Some mammals, all birds, and all reptiles lay amniotic eggs. **True.**

Do marsupials give birth to full term young or immature young? **The marsupial young are born at an immature stage.**

What is a joey? **A baby marsupial.**

What is a placenta? **It is the organ that forms between mother and developing baby to provide nutrition and waste disposal for the baby.**

16.4 MAMMAL TYPES

- A pictorial description of animal groups is presented.

Topic questions:

True or False? An elephant is a type of ungulate. **False. Ungulates are animals with hooves and elephants do not have them.**

True or False? The animals in Edentata have no teeth but do have other specialized structures to help them obtain their food. **True.**

What are the only mammals that can fly? **The bats—order Chiroptera.**

What are pinnipeds commonly known as? **Walruses, sea lions, and seals.**

True or False? Primates are the only organisms to have nails instead of claws. **True.**

16.5 KEY CHAPTER POINTS

- All birds lay eggs and share the following characteristics: feathers, beaks, light-weight skeleton, air sacs, and endothermic systems.
- The feather is a marvelously designed structure.
- Birds have especially large cerebellums, which is the area of the brain that controls balance and coordination.
- Air sacs allow birds to perform gas exchange when they inhale and exhale.
- The environment a bird lives in and the type of food it eats can be identified by looking at its beak and its feet.
- There are three categories of mammals—monotremes, marsupials, and placental mammals.
- Placental mammals are further divided into:
    - rodents (Rodentia)
    - insect-eating mammals (Insectivora)
    - toothless mammals (Edentata)
    - flying mammals (Chiroptera)
    - hoofed mammals (Artiodactyla and Perissodactyla)
    - trunk-nosed mammals (Proboscidea)
    - carnivores (Carnivora)
    - aquatic mammals (Cetacea and Sirenia)
    - primates (monkeys, apes, and humans)

17 | Human Anatomy and Physiology:
Control, Support, and Movement

17.0 CHAPTER PREVIEW

In this chapter we will:

- Investigate the structure and function of the nervous system, with attention to:
    - the individual nerve cell (neuron)
    - motor neurons, sensory neurons, and interneurons
    - the central and peripheral nervous systems
    - the connection between the central nervous and peripheral nervous systems
    - reflexes

- Describe how our special senses of hearing, taste, and vision are structured and function.

- Describe the anatomy of the endocrine system.

- Discuss how the nervous and endocrine systems interact to chemically control the body.

- Learn how the endocrine system works by studying examples of the endocrine diseases diabetes and hypothyroidism.

- Study the structure and function of the integumentary system (skin).

- Discuss the structure and function of the three types of muscles—skeletal, cardiac, and smooth.

- Discuss how muscles attach to bones and move joints.

- Learn the structure and function of bone.

- Review the function of ligaments and tendons.

- Investigate joint types.

17.1 OVERVIEW

- An overview of the organ systems we will cover is presented.

Topic questions: none

17.2 THE SUPER-CONTROLLER: THE NERVOUS SYSTEM

- The nervous system is a complex collection of nerves that includes the brain, spinal cord, and peripheral nerves.

- The nervous system controls everything that occurs in our bodies—movement, when we get hungry, how much salt we should retain, and when we need to sleep.

Topic question:

True or False? In order to control the bodily functions, the nervous system needs to be able to communicate with all tissues. **True.**

17.3 NERVOUS SYSTEM: NEURONS

- The nervous system is made from a nerve cell, called a neuron.

Topic questions:

True or False? Neurons are designed to be able to send and receive electrical impulses. **True.**

What are the three basic parts of a neuron? **The cell body, axon, and dendrites.**

17.4 NERVOUS SYSTEM: NEURON TYPES

- There are three basic neuron types—sensory, motor, and interneurons.

Topic questions:

True or False? Sensory neurons carry information from the spinal cord to the muscles. **False. Sensory neurons carry information from the skin and other organs to the spinal cord and brain.**

What does an interneuron do? **It carries information from one nerve to another.**

17.5 ORGANIZATION OF THE NERVOUS SYSTEM

- In the body (arms, legs, and trunk), sensory and motor neurons combine to form peripheral nerves. This collection of peripheral nerves is called the peripheral nervous system.

- In the middle of the body, sensory, motor, and interneurons combine to form the brain and spinal cord. This collection of neurons is called the central nervous system.

Topic questions:

True or False? The two basic components of the nervous system are the peripheral and the central nervous system. **True.**

What are the meninges? **They are the tissue that covers the brain and spinal cord to help protect them.**

17.6 CNS: BRAIN

- The brain is the organ that receives and processes all the information that is brought to it by the peripheral nerves and special sense organs.

Topic question:

True or False? The brain is not structured so an area of the brain controls specific actions or movements and different areas of the brain have different functions in all people. **False. The brain functions have been well-mapped and are consistent from person to person.**

17.7 CNS: SPINAL CORD

- The spinal cord is the pathway that carries information to and form the brain.

Topic question:

True or False? Unlike the brain, the white matter of the spinal cord is located near the outside and the gray matter is located on the inside. **True.**

17.8 PERIPHERAL NERVOUS SYSTEM

- The peripheral nervous system is the connection between the muscles and skin and the spinal cord. It is a collection of motor and sensory neurons that is specialized to carry information to and from the spinal cord.

Topic question:

What type of information do peripheral nerves carry and where does it go? **The peripheral nerves carry information for the muscles to contract or relax from the spinal cord to muscles. They also carry sensory information from muscles and skin to the spinal cord.**

17.9 REFLEXES

- Reflexes are "built in" connections between sensory nerves, the spinal cord, and motor nerves. They are for protection, so when we are in danger, we can respond quickly.

Topic question:

What is a reflex arc and what does it consist of? **Reflexes are "built in" connections between sensory nerves, the spinal cord, and motor nerves. A reflex arc consists of a sensory nerve, an interneuron in the spinal cord, and a motor nerve.**

17.10 SPECIAL SENSES: HEARING

- The ear is a structure that helps us to hear and to balance.

Topic question:

Describe the mechanism of hearing. **Sound enters the auditory canal and hits the eardrum. The sound causes the eardrum to vibrate. When the eardrum vibrates, so do the ear bones. The attachment of the middle ear bones to the cochlea causes a membrane on the cochlea to vibrate. The fluid in the cochlea moves when the membrane is vibrated by the moving of the middle ear bones. The movement of the fluid vibrates tiny hairs lining the cochlea. These hairs are connected to the nerves that transmit sound to the brain. Whenever the hairs move, information is sent to the temporal lobes of the brain, and the sound is interpreted.**

17.11 SPECIAL SENSES: VISION

- Eyes allow us to see.

Topic question:

Does a beam of light pass through the cornea or the retina first as it passes from the front of the eye to the back of the eye? **It passes through the cornea first.**

17.12 SPECIAL SENSES: TASTE AND SMELL

- Taste and smell function in a similar way.

Topic questions:

True or False? Certain areas of the tongue are more sensitive to tastes than others. **True.**

True or False? We can smell things because the chemical vapors diffuse into the brain and stimulate the brain tissue. **False. We have receptors in our sinuses that receive certain smell molecules. When a molecule fits into a receptor, we smell that smell.**

17.13 ENDOCRINE SYSTEM: GENERAL

- The endocrine system is a system that works together with the nervous system to provide chemical control over other organs and tissues.

Topic question:

What is a hormone? **Hormones are chemicals made in one tissue that are released into the blood stream and circulate to other tissues where they cause an effect on that organ's function.**

17.14 ENDOCRINE SYSTEM: REGULATION

- The hypothalamus is the controller of the endocrine system and works through the pituitary gland. Both are structures in the brain.

Topic question:

Describe the way that the endocrine system is controlled. **If an organ is not as active as it should be, the hypothalamus increases its activity by telling the pituitary gland to release a stimulating hormone. The stimulating hormone causes the activity of the target organ to increase. If an organ is over-active, then the hypothalamus tells the pituitary to stop releasing the stimulating hormone, so the target organ can shut down for a while.**

17.15 ENDOCRINE ABNORMALITIES

- Diabetes mellitus type I and hypothyroidism are examples of endocrine abnormalities.

Topic question:

True or False? People with type I diabetes produce too much insulin. **False. They do not produce any insulin.**

17.16 SKIN, MUSCLES, AND BONES

- Muscles are in the muscular system.
- Bones form the skeletal system.
- Skin is the integumentary tissue system.

Topic question:

True or False? Ligaments connect muscle to bones. **False. Tendons connect muscles to bones.**

17.17 SKIN

- The skin consists of two layers, the dermis and epidermis.

Topic question:

What is keratin? **It is the protein that makes up skin cells and gives skin its toughness.**

17.18 MUSCLES

- There are three types of muscle cells—striated (skeletal), cardiac, and smooth.

Topic question:

What is the difference between skeletal and smooth muscle? **Skeletal muscle is under active control, smooth muscle is not. Skeletal muscle has a banded appearance when viewed under the microscope, smooth muscle does not.**

17.19 BONES

- Bones make up the skeletal system, or skeleton.
- Joints form wherever two bones meet.

Topic questions:

What is a bone cell called? **An osteocyte.**

What is the difference between a fixed and moveable joint? **Fixed joints do not move and moveable joints do.**

17.20 KEY CHAPTER POINTS

- The nervous system is the super controller of the body.
- Humans are unique mainly because of the structure and function of our brains.
- The basic functional unit of the nervous system is the neuron.
- The nervous system components are the brain, spinal cord, and peripheral nerves.
- The nervous system is divided into the central nervous system (CNS) and the peripheral nervous system (PNS).
- The CNS includes the brain and the spinal cord.
- The PNS includes the motor and sensory nerves.
- Ears, eyes, nose, and tongue function as special extensions of the nervous system.
- The endocrine system functions with the nervous system to chemically control bodily functions through hormones.
- Diabetes mellitus type I and hypothyroidism are examples of endocrine diseases.
- Skin is the protective covering of the body; it is formed in two layers—the dermis and epidermis.
- There are three types of muscle cells—skeletal, smooth, and cardiac. Their function is to move some part of the body in some way.
- Tendons attach muscles to bones.

- Muscles cross at least one joint and move the joint they cross.
- The function of the skeletal system is to keep us upright and protect vital organs.
- Bone is composed of an outer layer, called compact bone, and an inner layer, called spongy bone.
- Ligaments hold bones together where they meet to form joints.
- There are many different types of joints in the human body.

Nutrition, Circulation, Respiration, and Protection

18

18.0 CHAPTER PREVIEW

In this chapter we will:

- Discuss the components of the digestive system.
- Investigate the function of the digestive system with attention to:
    - mechanical digestion
    - chemical digestion
    - peristalsis
    - villi, microvilli, and absorption
- Discuss the components and function of the circulatory system.
- Review a two-loop circulatory system.
- Examine the internal heart anatomy with attention to the valves.
- Investigate how the heartbeat and blood pressure are generated.
- Study the structure and function of blood.
- Discuss the components of the respiratory system.
- Learn the function of alveoli.
- Investigate how inhalation and exhalation occur.
- Discuss the components and function of the excretory system.
- Investigate how the kidney filters waste from blood and spares nutrients, ions, and water.
- Study the function of the immune system.
- Define pathogen.
- Discuss autoimmune diseases and organ transplantation as they relate to the immune system.

18.1 DIGESTIVE SYSTEM: GENERAL

- The function of the digestive system is to break down and absorb the substances needed to sustain life.

Topic question:

What are the substances we need to survive that our digestive system absorbs? **Nutrients.**

18.2 DIGESTIVE SYSTEM: COMPONENTS

- The components of the digestive tract are teeth, saliva, pharynx, esophagus, stomach, intestines, liver, pancreas, and gall bladder.

Topic question:

What is another name for the digestive system? **The GI tract.**

18.3 DIGESTIVE SYSTEM: FUNCTION

- The functions of the components of the digestive system are discussed.

Topic questions:

True or False? Saliva provides mechanical digestion. **False. Saliva provides chemical digestion.**

Where does the food go after the stomach has broken it into chyme? **The small intestine.**

Which structures are larger, villi or microvilli? **Villi.**

18.4 CIRCULATORY SYSTEM: GENERAL

- The purpose of the **circulatory system** is to move oxygen, hormones, and nutrients to tissues/organs and to move carbon dioxide and wastes from tissues/organs.

Topic question:

What is blood with low oxygen concentration called? **Deoxygenated blood.**

18.5 CIRCULATORY SYSTEM: THE HEART

- The heart has four chambers and four valves.

Topic question:

What are the names of the heart valves? **Tricuspid, pulmonic, mitral, and aortic.**

18.6 CIRCULATORY SYSTEM: THE HEARTBEAT AND BLOOD PRESSURE

- The impulse for the heartbeat comes from the heart itself, an electrical impulse that starts in the right atrium, in a bundle of tissue called the sinoatrial node, or S-A node.
- Blood pressure is the pressure that forms in the blood vessels as a result of the pumping action of the heart.

Topic question:

True or False? The right ventricle generates more pressure when it contracts than the left ventricle. **False.**

18.7 BLOOD

- Blood is an inseparable part of the circulatory system, but it is considered a connective tissue.
- Blood is a mixture of liquid—called plasma—in which blood cells are suspended.

Topic question:

True or False? Blood is composed of red blood cells, white blood cells, and plasma. **True.**

18.8 RESPIRATORY SYSTEM: COMPONENTS

- The purpose of the respiratory system is to bring fresh, oxygen-containing air into the lungs, so gas exchange can occur. Then, the air from the lungs, filled with carbon dioxide, needs to be expelled.

Topic question:

What occurs in an alveolus? **Gas exchange.**

18.9 RESPIRATORY SYSTEM: FUNCTION

- Inhalation occurs through the muscular contraction of the diaphragm and exhalation occurs through the natural deflating action of the lung tissue.

Topic question:

What is the phrenic nerve? **It is the nerve that carries the impulse to breathe from the brain to the diaphragm.**

18.10 EXCRETORY SYSTEM

- The excretory system's job is to filter the blood and eliminate wastes, store them for a short time, then eliminate them from the body.

Topic question:

All the following are parts of the excretory system except: nephron, intron, kidney, or ureters. **Intron.**

18.11 IMMUNE SYSTEM: OVERVIEW

- The immune system protects from infections and fights them when we contract one.

Topic question:

What is a pathogen? **It is an organism that can cause a disease (or infection).**

18.12 IMMUNE SYSTEM: FUNCTION

- The immunologic system has two ways that it fights infection—keep the organism out (the barrier function)—and fight the organism once it enters (the active function).

Topic questions:

What happens when the barrier function of the immune system fails? **The active function of the immune system kicks in and we start to actively fight the pathogen.**

What do antibodies do? **They stick to and coat pathogens, targeting them for destruction by cells of the immune system.**

18.13 IMMUNE SYSTEM: ACTIVE IMMUNITY

- Following an infection, our immune system holds onto the ability to recognize the same pathogen forever. This is called active immunity.

Topic question:

True or False? Passive immunity develops after we fight off an infection. **False. Active immunity develops after an infection or an immunization.**

18.14 IMMUNE SYSTEM: AUTOIMMUNE RESPONSE

- Sometimes, for some reason, the immune system is stimulated to attack the body. This is called an autoimmune disease.

Topic question:

Name two conditions that prompt doctors to purposely prescribe medications to suppress a person's immune system so it cannot work well. **After organ transplantation and autoimmune disease.**

18.15 KEY CHAPTER POINTS

- The digestive system breaks down and absorbs nutrients from the food we eat.
- There are two parts of food breakdown—mechanical and chemical.
- Mechanical digestion occurs in the mouth and stomach.
- Chemical digestion occurs in the mouth, stomach, and first part of the small intestine.
- The salivary glands, liver, and pancreas also play an important role in the chemical digestion of food.
- Absorption occurs in the second and third parts of the small intestine.
- The large intestine packages waste for removal and absorbs water from the waste material before it is eliminated.
- The purpose of the circulatory system is to move oxygen, hormones, and nutrients to tissues/organs; and carbon dioxide and wastes from tissues/organs.
- Humans have a four-chambered heart and a two-loop circulatory system.
- The heart has four valves to prevent the backflow of blood.
- The electrical activity for the heartbeat originates in the SA node.
- Blood pressure is generated by the contraction of the ventricles.
- Blood is composed of plasma, red blood cells, and white blood cells.
- Red blood cells contain hemoglobin, which carries oxygen to the tissues.

- The functional unit of the lung is the alveolus.
- Inhalation occurs due to the contraction of the diaphragm.
- Exhalation occurs due to the relaxation of the diaphragm and the natural property of the lungs to deflate.
- The excretory system filters blood and eliminates wastes, stores them for a short time, then eliminates them from the body while sparing water, ions, and nutrients.
- The functional unit of the kidney is the nephron.
- The immune system is a complex system protecting us from infectious organisms.
- The immune system employs both physical barriers and targeted removal to rid the body of pathogens.
- Sometimes the immune system needs to be suppressed due to organ transplantation or autoimmune disease.

19 | Earth Studies

19.0 CHAPTER PREVIEW

In this chapter we will:

- Introduce basic concepts and terms of ecology.

- Discuss the concepts of the biosphere.

- Investigate ecosystems with attention to:
  - abiotic mass
  - biotic mass
  - water, carbon, and oxygen cycles
  - pollution
  - energy transfer within predator-prey relationships
  - camouflage and mimicry

- Discuss marine and fresh water biomes.

- Discuss the seven terrestrial biomes.

19.1 OVERVIEW

- Ecology is a word that was coined in the late 1800s to describe the study of the relationships living organisms have with one another and their physical environment.

- Ecologists are the scientists who study ecology.

Topic question:

What is an ecologist? **A scientist who studies ecology.**

19.2 BIOSPHERE

- The biosphere is the area of earth that can support life.

Topic question:

How "thick" is the biosphere? **The biosphere encompasses an area from about six miles above sea level to seven miles below sea level, so the total area is thirteen miles.**

19.3 ECOSYSTEMS

- An ecosystem is the association and interaction of all living organisms within their physical environment.

Topic question:

How are ecosystems defined? **The boundaries are defined by the researcher who is studying it.**

19.4 COMPONENTS OF ECOSYSTEMS

- There are two general components to every ecosystem—biotic and abiotic mass.

Topic question:

What is the difference between biotic and abiotic mass in an ecosystem? **Biotic mass is defined as all living organisms in an ecosystem. Abiotic mass is everything in the ecosystem that is not alive.**

19.5 ABIOTIC COMPONENTS OF ECOSYSTEMS

- Abiotic components like rocks, water, soil, sand, and man-made objects. Also, abiotic components include temperature, soil conditions, and the cycles that describe the flow of water and elements between abiotic and biotic mass.

Topic questions:

True or False? Abiotic components of an ecosystem can include the water cycle, a rock, soil conditions, temperature, and the amount of sunlight an ecosystem receives. **True.**

What are the similarities and differences between percolation, precipitation, evaporation, and transpiration? **These are all terms that relate to movements of water in the water cycle. Percolation is the movement of water down into the ground. Precipitation is water that falls from the sky to the ground (the water can be in any form). Evaporation is water that turns into a gas (water vapor) from a body of water. The water vapor then returns to the atmosphere. Transpiration is the process of plants losing water into the atmosphere when they are photosynthesizing. This is kind of like evaporation from plants.**

19.6 BIOTIC COMPONENTS OF ECOSYSTEMS

- Biotic components of ecosystems consist of everything that is alive in the ecosystem. This includes organisms seen and unseen.

Topic question:

True or False? The biotic components of an ecosystem are usually grouped into communities, populations, and individuals. **True.**

19.7 ENERGY IN ECOSYSTEMS

- There are three common methods of showing the energy transfer relationships in an ecosystem: food chains, food pyramids, and food webs.

Topic question:

Which is the most accurate way to depict the energy transfer relationships that exist in ecosystems? **A food web.**

19.8 ANIMAL RELATIONSHIPS IN ECOSYSTEMS

- Animal relationships include predator-prey relationships.
- Prey species have special physical traits that make them more effective at eluding being caught and eaten.
- Predator species have special features that make them more effective hunters.

Topic questions:

What is the condition called when a prey species is shaped like a leaf? **Camouflage.**

What is the term that describes a harmless species resembling a harmful one? **Mimicry.**

19.9 BIOMES

- A **biome** is a large area of earth that contains similar flora and fauna with a physical environment that looks similar over the entire area.

Topic question:

What is a large area of the earth that has similar temperatures, soil conditions, and plant and animal life called? **A biome.**

19.10 PIONEER SPECIES AND CLIMAX COMMUNITIES

- When areas that are not populated by plant or animal life begin to be populated, there is a similarly observed process of population into the areas.

Topic question:

What is the process an unpopulated area of land undergoes to become an ecosystem? **The first species that move in are called pioneer species. These are plants that are usually weeds, lichens, and other species able to live in nutrient-poor soil. Pioneer species begin to loosen up the soil and provide increasing organic matter. The first plants and animals into the area form primary succession. When more fertile soil is present, larger species of plants and animals move in during the process of secondary succession. When the flora and fauna of an ecosystem are stable, then the climax community has been reached.**

19.11 KEY CHAPTER POINTS

- The biosphere describes the total area of the earth that can be inhabited by organisms.
- An ecosystem is the association and interaction of all living organisms within their physical environment.
- Ecosystems are composed of biotic and abiotic components.
- Abiotic components include non-living physical substances in an ecosystem as well as temperature, sunlight, and soil conditions.
- Biotic components are all bacterial, protist, algae, plant, and animal life in an ecosystem.
- The transfer of energy in an ecosystem can be depicted using a food web.
- Predator-prey relationships and consumer-producer relationships are demonstrated in food webs.
- Predators and prey have specialized behaviors and physical traits—such as herding, camouflage, and mimicry—which make them more effective at surviving.
- A biome is a large area of earth that contains similar flora and fauna with a physical environment that looks similar over the entire area.
- There are fresh water and marine aquatic biomes as well as seven terrestrial biomes.
- A climax community is an ecosystem that has reached stability in regard to the flora and fauna living there.